6.88

STUDENT SUPPLEMENT

Volume 2: Chapters 13-19

to accompany

Swokowski's CALCULUS With Analytic Geometry
Second Edition

by

Thomas A. Bronikowski
Marquette University

Prindle, Weber & Schmidt
Boston, Massachusetts

PWS PUBLISHERS

Prindle, Weber & Schmidt · Willard Grant Press · Duxbury Press ·
Statler Office Building · 20 Providence Street · Boston, Massachusetts 02116

Printed in the United States of America.
10 9 8 7 6 5 — 86 85 84 83 82

ISBN 0-87150-278-X

PWS Publishers is a division of Wadsworth, Inc.

INTRODUCTION

This supplement to the second edition of Earl W. Swokowski's text, Calculus with Analytic Geometry, has been written to help you, the student, develop the skills and art of solving Calculus problems. In it I have included the solution of every third problem (numbers 1, 4, 7, 10, etc.) from every section of the text. Very occasionally this scheme is slightly modified to provide a greater variety of exercises and to furnish hints for some problems not in the usual sequence. There is a special feature of this second edition which should be of significant help to you in sections of the text concerned with applications of the integral. (These are principally in Chapters 6 and 10 of Volume 1 and Chapters 13, 15 and 17 of Volume 2.) In addition to the complete solutions of the usual exercises, I have included in such sections the integral forms of the answers to the remaining odd-numbered exercises (numbers 3, 5, 9, 11, etc.). In this way you can check the "set-up" of the solution of the exercise before wasting precious time evaluating the wrong integral.

Obviously not every arithmetic and algebraic detail can be included in every solution if the size of this work is to remain reasonable. You will find sufficient detail in every solution to see how the answer is obtained. As a rule, more calculational details are shown in the first problem of a set than in the last problems of that set. The symbols " $\Longrightarrow$ " and " $\Longleftrightarrow$ " are frequently used to denote implication and equivalence, respectively, between the mathematical expressions they connect. All other symbols will be familiar to you from the text or from previous courses. The author would appreciate being informed of any errors or misprints that you find. Just send a note to me at the Department of Mathematics and Statistics, Marquette University, Milwaukee, WI, 53233.

I wish to thank all of those who have spoken or written to me pointing out errors and suggesting improvements in the first edition. I would also like to express my appreciation to the staff of Prindle, Weber and Schmidt, especially John Martindale and David Chelton, for their assistance and advice, to my friend and colleague, Professor Earl W. Swokowski, for his thoughtful recommendations, and to Sally Canapa for her excellent job of typing the entire manuscript. To my wife, Irene, and our children, Joseph, Michael and Anne, I owe a special debt of gratitude for their encouragement and assistance in the preparation of this edition. Particular appreciation is expressed to the boys for their help in proofreading and in making several new sketches and to my wife for her generous and unselfish assistance in proofreading and in preparing the final version of the finished manuscript.

Thomas A. Bronikowski
Department of Mathematics and Statistics
Marquette University
Milwaukee, WI 53233

TABLE OF CONTENTS

Chapter 13	Plane Curves and Polar Coordinates	1
Chapter 14	Vectors and Solid Analytic Geometry	20
Chapter 15	Vector-Valued Functions	34
Chapter 16	Partial Differentiation	43
Chapter 17	Multiple Integrals	62
Chapter 18	Topics in Vector Calculus	96
Chapter 19	Differential Equations	120

CHAPTER 13

PLANE CURVES AND POLAR COORDINATES

EXERCISES 13.1, page 615

1. Solving the 1st equation for t gives $t = x+2$. Substituting this into the 2nd equation we get $y = 2(x+2)+3 = 2x+7$. When $t = 0$, $x = -2$ and when $t = 5$, $x = 3$. Thus the limits $-2 \leq x \leq 3$.

4. Subtracting the 2 equations yields $x-y = 2$ or $y = x-2$, for $-7 \leq x \leq 9$. (When $t = -2$, $x = -7$, and when $t = 2$, $x = 9$.)

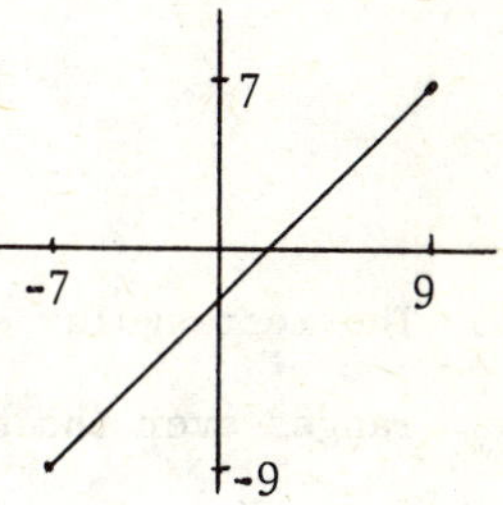

7. $x = e^t \Longrightarrow e^{-t} = 1/x$. $y = e^{-2t} = (e^{-t})^2 \Longrightarrow y = (\frac{1}{x})^2$.

10. The parameter can be eliminated by writing $\cos t = x+2$, $\sin t = y-3$. Then $1 = \sin^2 t + \cos^2 t = (y-3)^2 + (x+2)^2$, a circle of radius 1, C(-2,3).

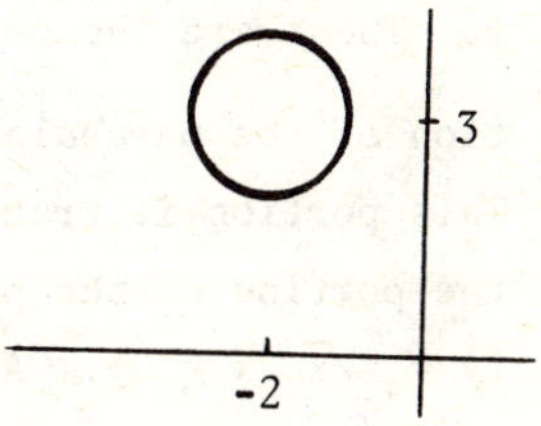

13. Since $t > 0$, $x = t^2 \Longrightarrow t = \sqrt{x} \Longrightarrow y = 2 \ln\sqrt{x} = \dfrac{2 \ln x}{2} = \ln x$.

16. $e^{-t} = 1/e^t \Longrightarrow y = 1/x$. But since $x = e^t > 0$, the graph is the right branch only.

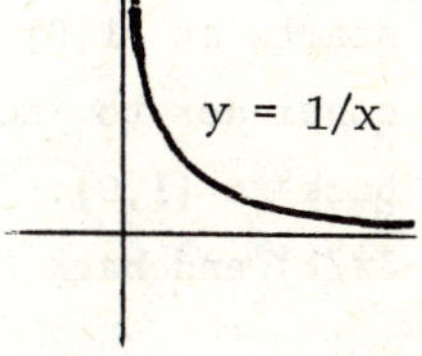

19. Substituting x for t, we get $y = \sqrt{x^2-1} \Longrightarrow y^2 = x^2-1 \Longrightarrow x^2-y^2 = 1$, a familiar hyperbola. Since $y \geq 0$, the portion of this hyperbola below the x-axis is not contained as part of C. Since $|x| = |t| \geq 1$, C extends to the right and left without limit.

22. $y = 8t^3 = (2t)^3 = x^3$, and $-1 \leq t \leq 1 \Rightarrow -2 \leq 2t = x \leq 2$ yield the graph shown. The graph is smooth since $\frac{dx}{dt} = 2$ and $\frac{dy}{dt} = 24t^2$ are continuous for $-1 \leq t \leq 1$ and are not both 0 there.

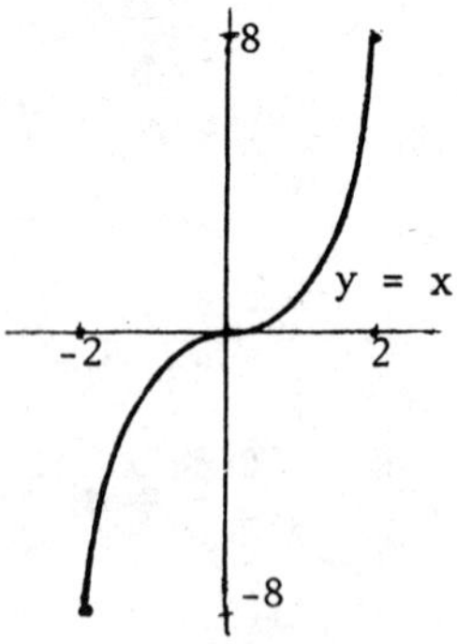

25. The rectangular equation is $x = y^2$. C_1 is the entire parabola since $y = t$ ranges over the entire set of real numbers. C_2 is the upper portion only ($y = \sqrt{x}$, $x \geq 0$) since $y = t^2 \geq 0$ here. As t ranges from $-\infty$ to ∞, this portion of the parabola is traced out twice. For C_3, note that on any t-interval of length 2π, $x = \sin^2 t$ ranges from 0 to 1, and y ranges from -1 to 1. These are the only possible values of x and y here. Thus C_3 is the portion of the parabola from the point (1,1) to (1,-1) through the vertex (0,0). This portion is traced out infinitely often as t goes from $-\infty$ to ∞. C_4 is the portion of the parabola below the x-axis not including the origin ($y = -\sqrt{x}$, $x > 0$). As $t \to -\infty$, the origin is approached, and as $t \to \infty$, the point moves endlessly to the right along this part of the parabola.

28. In each case the motion is on the line $y = 1-x$. (For part (c), $y = \sin^2 t = 1 - \cos^2 t = 1-x$.) (a) As t increases from 0 to 1, P starts at (0,1) and moves to (1,0). (b) Recalling that $\ln 1 = 0$ and $\ln e = 1$, as t increases from 1 to e, P starts at (1,0) and moves to (0,1) along this line. (c) P starts at (1,0) when $t = 0$. As t increases to $\pi/2$, P moves to (0,1). As t continues to increase to π, P reverses its direction of motion and moves back to (1,0). As t increases from π to 2π, P again moves to (0,1) (at $t = 3\pi/2$) and back, finally, to its starting point (1,0).

31. Solving each equation we get $\cos t = \frac{x-h}{a}$, $\sin t = \frac{y-k}{b}$ and thus

$$\frac{(x-h)^2}{a^2} + \frac{(y-k)^2}{b^2} = 1.$$

34. Let θ = angle FDP and α = angle ADP = angle EDP. Then angle ODA = $\frac{\pi}{2}$ - t and $\alpha = \theta - (\frac{\pi}{2} - t) = \theta + t - \frac{\pi}{2}$. Arcs GF and PF are equal in length since each is the distance rolled. Thus at = $b\theta$ or $\theta = \frac{a}{b}t$, and $\alpha = \frac{a+b}{b}t - \frac{\pi}{2}$. Note that $\cos\alpha = \sin\frac{a+b}{b}t$ and $\sin\alpha = -\cos\frac{a+b}{b}t$. For the location of the points as illustrated, the coordinates of P are:

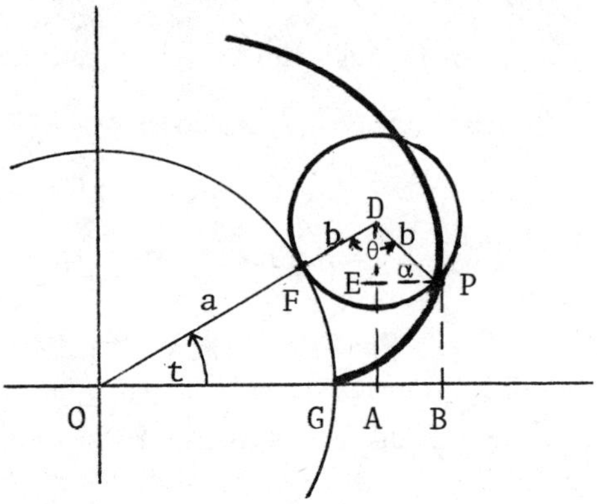

$$x = d(O,A) + d(A,B) = d(O,A) + d(E,P)$$
$$= (a+b)\cos t + b\sin\alpha = (a+b)\cos t - b\cos\frac{a+b}{b}t$$
$$y = d(B,P) = d(A,D) - d(D,E) = (a+b)\sin t - b\cos\alpha$$
$$= (a+b)\sin t - b\sin\frac{a+b}{b}t .$$

It can be verified that these equations are valid for all locations of the points.

EXERCISES 13.2, page 619

1. $x = t-2 \Rightarrow \frac{dx}{dt} = 1$, $y = 2t+3 \Rightarrow \frac{dy}{dt} = 2$. Thus $\frac{dy}{dx} = \frac{2}{1} = 2$ for all t. (Not very surprising since the graph is a line with slope 2.)

4. $x = t^3 + 1 \Rightarrow \frac{dx}{dt} = 3t^2$, $y = t^3 - 1 \Rightarrow \frac{dy}{dt} = 3t^2$. Thus $\frac{dy}{dx} = \frac{3t^2}{3t^2} = 1$ for all t.

7. $x = e^t \Rightarrow \frac{dx}{dt} = e^t$, $y = e^{-2t} \Rightarrow \frac{dy}{dt} = -2e^{-2t}$. Thus $\frac{dy}{dx} = \frac{-2e^{-2t}}{e^t} = -2e^{-3t}$. When t = 1, $\frac{dy}{dx} = -2e^{-3}$.

10. $x = \cos t - 2 \Rightarrow \frac{dx}{dt} = -\sin t$, $y = \sin t + 3 \Rightarrow \frac{dy}{dt} = \cos t$. Thus $\frac{dy}{dx} = \frac{\cos t}{-\sin t} = -\cot t$. When t = 1, $\frac{dy}{dx} = -\cot 1$.

13. Let a be the value of t at the unknown point of tangency. The slope of the tangent line when t = a is

$\frac{dy}{dx} = \left.\frac{dy/dt}{dx/dt}\right|_{t=a} = \frac{2}{8a} = \frac{1}{4a}$, if $a \neq 0$, and the equation of the tangent line is

$y-1 = \frac{1}{4a}(x-4)$. Since the point of tangency, $(4a^2, 2a-5)$, is on this line, its coordinates satisfy this equation. Substituting, we obtain the following equation for a: $(2a-5)-1 = \frac{1}{4a}(4a^2-4) = \frac{1}{a}(a^2-1) \iff 2a^2 - 6a = a^2-1 \iff a^2 - 6a + 1 = 0$. Solving by the quadratic formula we obtain $a = \frac{6 \pm \sqrt{36-4}}{2} = 3 \pm \sqrt{8} = 3 \pm 2\sqrt{2}$. Thus there are 2 tangent lines: $y-1 = \frac{1}{4(3\pm2\sqrt{2})}(x-4)$.

If a = 0, the tangent line is vertical and passes through (0,-5) and thus has equation x = 0 which does not pass through (4,1). Thus the 2 lines found above are the only solutions.

16. $x = t^3+1 \Rightarrow \frac{dx}{dt} = 3t^2$, $y = t^2-2t \Rightarrow \frac{dy}{dt} = 2t-2$.

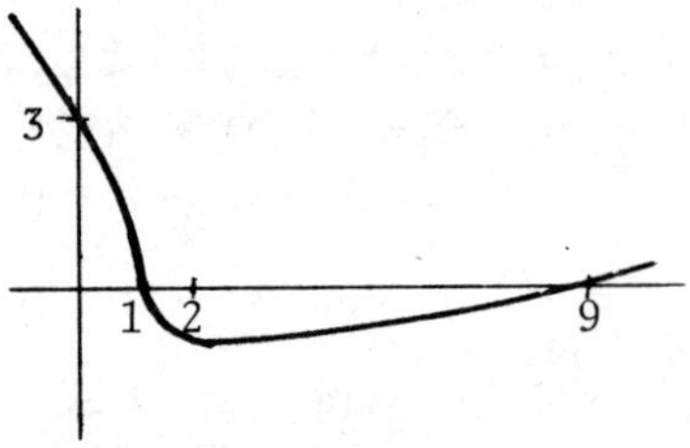

Thus $\frac{dy}{dx} = \frac{2t-2}{3t^2}$. This slope is 0 when t = 1. Thus there is a horizontal tangent line at (2,-1). The slope is undefined when t = 0 and becomes infinite as $t \to 0$. Thus there is a vertical tangent line at (1,0).

Finally, $\frac{d^2y}{dx^2} = \frac{dy'/dt}{dx/dt} = \frac{D_t((2t-2)/3t^2)}{3t^2}$. Since $D_t(\frac{2t-2}{3t^2}) = \frac{2}{3} D_t(\frac{1}{t} - \frac{1}{t^2}) = \frac{2}{3}(-\frac{1}{t^2} + \frac{2}{t^3}) = \frac{2}{3t^3}(2-t)$, we obtain $\frac{d^2y}{dx^2} = \frac{2}{3t^3}(2-t) \times \frac{1}{3t^2} = \frac{2(2-t)}{9t^5}$.

19. $x = a(t - \sin t) \Rightarrow \frac{dx}{dt} = a(1 - \cos t)$, $y = a(1 - \cos t) \Rightarrow \frac{dy}{dt} = a \sin t$.

Thus $\frac{dy}{dx} = \frac{\sin t}{1 - \cos t} = \frac{1}{\tan(t/2)}$. This is 0 when sin t = 0 and $\cos t \neq 1$, i.e. when t is an odd multiple of π. The points are, therefore, $(a(2n+1)\pi, 2a)$. The tangent line will be vertical when cos t = 1, i.e. when t is an even multiple of π. (Notice that the numerator of $\frac{dy}{dx}$, sin t, is also 0 for such t. However, using L'Hôpital's rule $\lim_{t\to2n\pi} \frac{dy}{dx} = \lim_{t\to2n\pi} \frac{\cos t}{\sin t} = \pm \infty$.) The points at which the tangent line is vertical are thus $(a2n\pi, 0)$. The points at which the slope is 1 are most easily found by using the tan(t/2) formulation. Setting tan(t/2) = 1 we get $t/2 = \pi/4 + n\pi$ or $t = \pi/2 + 2n\pi$. The points are thus $(a(\pi/2 + 2n\pi - 1), a)$.

EXERCISES 13.3, page 625

1.
4. Both equations are of the form r = c, or $\sqrt{x^2+y^2} = c$, which becomes $x^2 + y^2 = c^2$. Thus the graphs are circles centered at the origin of radii 5 and 2, respectively. (The radius of $x^2 + y^2 = c^2$ is $|c|$.)

7. The graph shown in the text can be obtained by noting the changes described in the following table and then plotting points.

θ	$\sin\theta$	$r=4(1-\sin\theta)$
$0 \to \frac{\pi}{2}$	$0 \to 1$	$4 \to 0$
$\frac{\pi}{2} \to \pi$	$1 \to 0$	$0 \to 4$
$\pi \to \frac{3\pi}{2}$	$0 \to -1$	$4 \to 8$
$\frac{3\pi}{2} \to 2\pi$	$-1 \to 0$	$8 \to 4$

10. Again, we tabulate

θ	4θ	$r=2\sin 4\theta$
$0 \to \frac{\pi}{8}$	$0 \to \frac{\pi}{2}$	$0 \to 2$
$\frac{\pi}{8} \to \frac{\pi}{4}$	$\frac{\pi}{2} \to \pi$	$2 \to 0$
$\frac{\pi}{4} \to \frac{3\pi}{8}$	$\pi \to \frac{3\pi}{2}$	$0 \to -2$
$\frac{3\pi}{8} \to \frac{\pi}{2}$	$\frac{3\pi}{2} \to 2\pi$	$-2 \to 0$

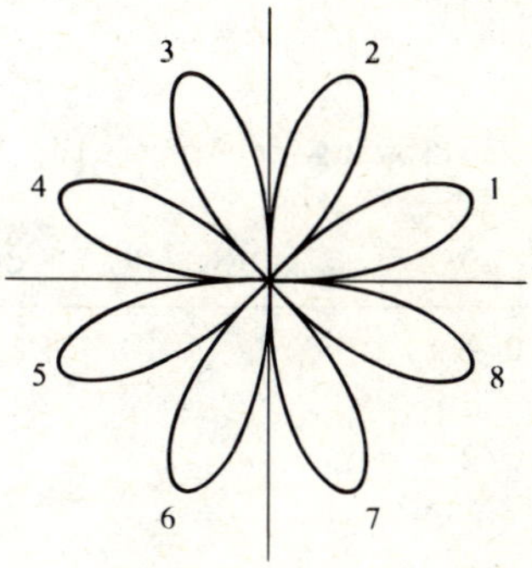

These calculations produce leaves 1 and 6 (counting counterclockwise from the +x-axis) of the 8-leaved rose. As θ continues increasing to 2π, the pattern seen in the table continues, producing the other 6 leaves in the order 3, 8, 5, 2, 7, 4.

13. The graph shown in the text is most easily obtained by writing the equation in rectangular form. $r = 4\csc\theta = 4/\sin\theta \Longrightarrow r\sin\theta = 4 \Longrightarrow y = 4$.

16. We first note that the graph of $r = 2\sec\theta$ is the vertical line $x = 2$ since $r = 2\sec\theta \iff r = \dfrac{2}{\cos\theta} \iff r\cos\theta = 2 \iff x = 2$. The given equation

is $r = 2 + 2 \sec \theta$, and for $-\frac{\pi}{2} < \theta < \frac{\pi}{2}$ (where $\sec \theta > 0$), the point on the graph corresponding to an angle θ_o in this interval is 2 units beyond (i.e. to the right of) the line $x = 2$. This produces the right branch of the curve as illustrated. Similarly if $\frac{\pi}{2} < \theta < \frac{3\pi}{2}$ (where $\sec \theta \leq -1$) the point is 2 units to the left of the line measured on a ray $\theta = \theta_o$, where $\frac{\pi}{2} < \theta_o < \frac{3\pi}{2}$. This produces the left branch of the curve.

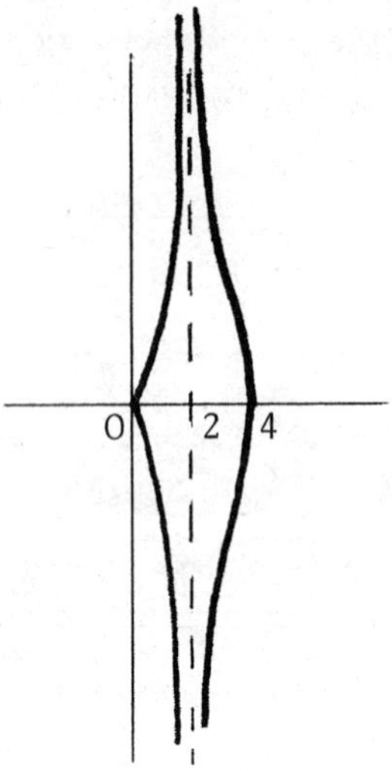

19. The following table yields the graph shown in the text.

θ	$\cos \theta$	$-6(1 + \cos \theta)$	
$0 \to \pi/2$	$1 \to 0$	$-12 \to -6$	lower half
$\pi/2 \to \pi$	$0 \to -1$	$-6 \to 0$	
$\pi \to 3\pi/2$	$-1 \to 0$	$0 \to -6$	upper half
$3\pi/2 \to 2\pi$	$0 \to 1$	$-6 \to -12$	

The first two lines of the table produce the lower half even though $0 \leq \theta \leq \pi$ since $r \leq 0$ for such θ, and similarly for the last two lines producing the upper half.

22. To get the correct placement of the five leaves and the order in which they are traced out, we again tabulate. The increments in θ, $\pi/10$, are determined by letting 5θ increase by $\pi/2$.

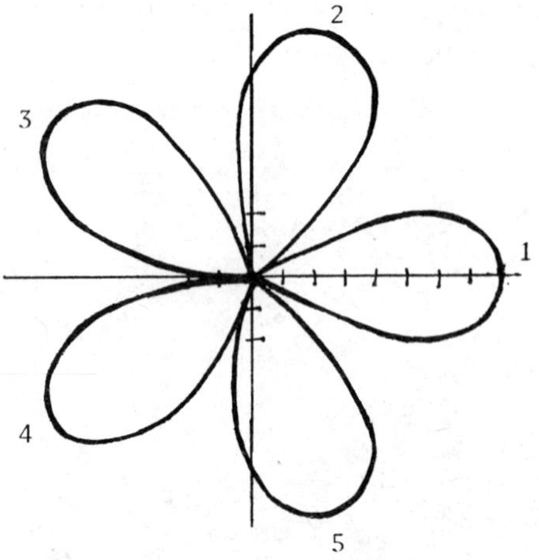

θ	5θ	$8\cos 5\theta$	Leaf
$0 \to \pi/10$	$0 \to \pi/2$	$8 \to 0$	1 (top half)
$\pi/10 \to 2\pi/10$	$\pi/2 \to \pi$	$0 \to -8$	4
$2\pi/10 \to 3\pi/10$	$\pi \to 3\pi/2$	$-8 \to 0$	
$3\pi/10 \to 4\pi/10$	$3\pi/2 \to 2\pi$	$0 \to 8$	2
$4\pi/10 \to 5\pi/10$	$2\pi \to 5\pi/2$	$8 \to 0$	
$5\pi/10 \to 6\pi/10$	$5\pi/2 \to 3\pi$	$0 \to -8$	5
$6\pi/10 \to 7\pi/10$	$3\pi \to 7\pi/2$	$-8 \to 0$	
$7\pi/10 \to 8\pi/10$	$7\pi/2 \to 4\pi$	$0 \to 8$	3
$8\pi/10 \to 9\pi/10$	$4\pi \to 9\pi/2$	$8 \to 0$	
$9\pi/10 \to \pi$	$9\pi/2 \to 5\pi$	$0 \to -8$	1 (bottom half).

25. $x = -3 \Longrightarrow r\cos\theta = -3 \Longrightarrow r = -3/\cos\theta = -3\sec\theta$.

28. $x^2 = 8y \Longrightarrow r^2\cos^2\theta = 8r\sin\theta \Longrightarrow r = 8\sin\theta/\cos^2\theta = 8\tan\theta\sec\theta$. (We assumed $r \neq 0$ in cancelling it even though the origin satisfied the original equation. There is no problem, though. The origin is still obtained as a solution by taking $\theta = 0$.)

31. $x^2 - y^2 = 16 \Longrightarrow r^2\cos^2\theta - r^2\sin^2\theta = 16 \Longrightarrow r^2\cos 2\theta = 16 \Longrightarrow r^2 = 16\sec 2\theta$.

34. $r\sin\theta = -2 \Longrightarrow y = -2$.

37. $r = 2 \Longrightarrow \pm\sqrt{x^2+y^2} = 2 \Longrightarrow x^2 + y^2 = 2^2 = 4$.

40. $r = 4\sec\theta \Longrightarrow x = 4$ (See #16).

43. $r^2\cos 2\theta = 1 \Longrightarrow r^2(\cos^2\theta - \sin^2\theta) = 1 \Longrightarrow r^2\cos^2\theta - r^2\sin^2\theta = 1 \Longrightarrow x^2-y^2 = 1$, a familiar hyperbola.

46. $r(\sin\theta + r\cos^2\theta) = 1 \Longrightarrow r\sin\theta + r^2\cos^2\theta = 1 \Longrightarrow y + x^2 = 1$, or $y = 1 - x^2$.

EXERCISES 13.4, page 632

1. $r = \frac{12}{6 + 2\sin\theta} = \frac{2}{1 + \frac{1}{3}\sin\theta} \Rightarrow e = \frac{1}{3}$, $de = 2$ or $d = 6$. Thus, an ellipse with major axis along the y-axis. (The presence of $\sin\theta$ tells us that.) The vertices occur when $\theta = \frac{\pi}{2}$ and $\frac{3\pi}{2}$ yielding $r = \frac{3}{2}$ and 3, respectively.

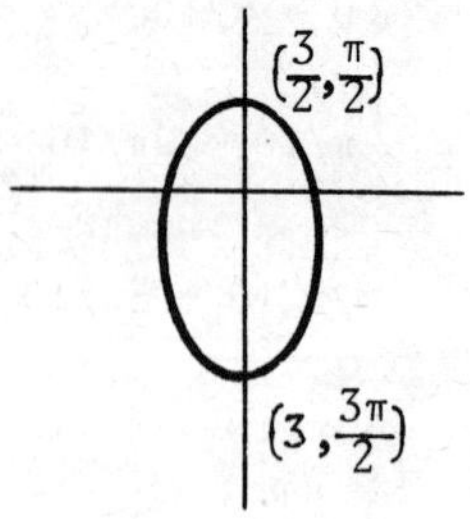

4. $r = \frac{12}{2 + 6\cos\theta} = \frac{6}{1 + 3\cos\theta} \Rightarrow e = 3$, $de = 6$ or $d = 2$. Thus, a hyperbola with transverse axis along the polar axis. The vertices occur when $\theta = 0$ and π yielding $r = 3/2$ and -3 respectively.

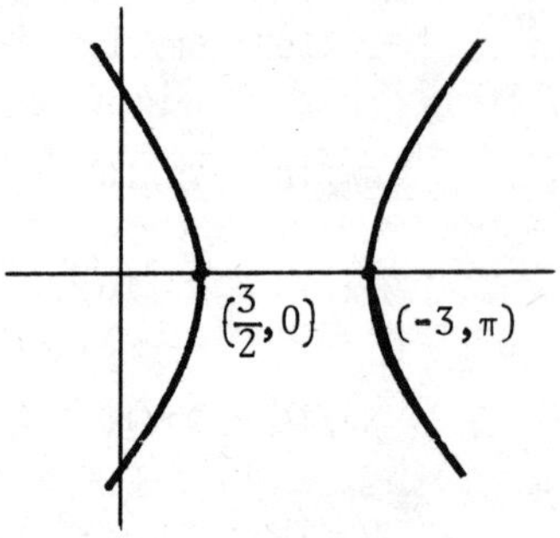

7. $r = \frac{4}{\cos\theta - 2} = \frac{-4}{2 - \cos\theta} = \frac{-2}{1 - \frac{1}{2}\cos\theta} \Rightarrow$ $e = \frac{1}{2}$ and the graph is an ellipse with major axis along the polar axis. The vertices occur when $\theta = 0$ and π yielding $(-4,0)$ and $(-4/3,\pi)$.

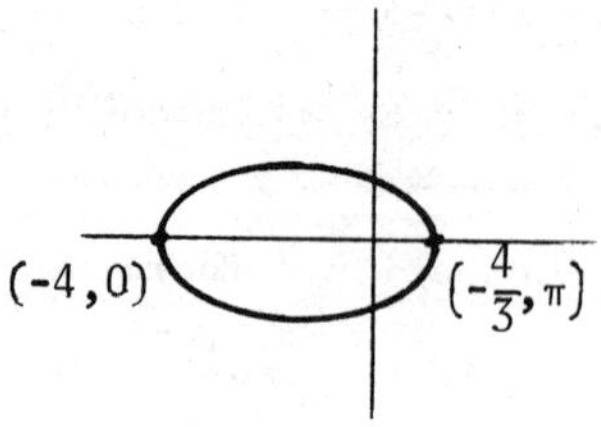

10. $r = \csc\theta(\csc\theta - \cot\theta) = \frac{1}{\sin^2\theta} - \frac{\cos\theta}{\sin^2\theta} =$ $\frac{1 - \cos\theta}{\sin^2\theta} = \frac{1 - \cos\theta}{1 - \cos^2\theta} = \frac{1}{1 + \cos\theta} \Rightarrow e = 1$ and the graph is almost a parabola with focus $(0,0)$, directrix the vertical line $r = \sec\theta$. Note that the vertex of this parabola, $(1/2,0)$, is not on the graph of the original equation since $\theta = 0$ is not in the domain of $\csc\theta$ or $\cot\theta$.

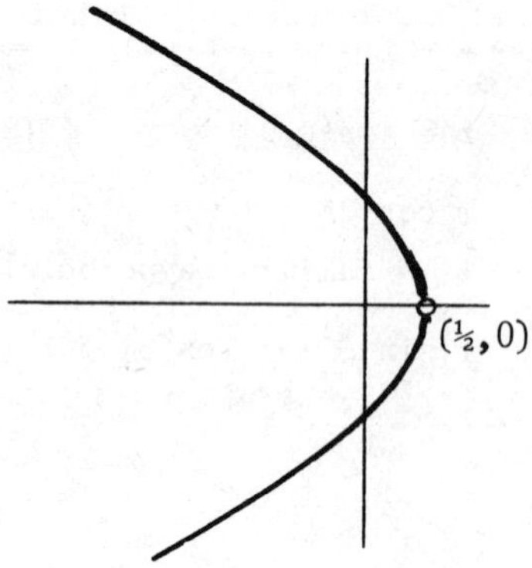

13. $r = \dfrac{12}{2 - 6\cos\theta} \Longrightarrow 2r - 6r\cos\theta = 12 \Longrightarrow \underline{+}\sqrt{x^2+y^2} = 3x+6 \Longrightarrow x^2 + y^2 =$ $(3x+6)^2 = 9x^2 + 36x + 36 \Longrightarrow y^2 - 8x^2 - 36x - 36 = 0.$

16. $r = \dfrac{3}{2 - 2\sin\theta} \Longrightarrow 2r - 2r\sin\theta = 3 \Longrightarrow \underline{+}2\sqrt{x^2+y^2} = 2y + 3 \Longrightarrow 4(x^2+y^2) =$ $4y^2 + 12y + 9 \Longrightarrow 4x^2 - 12y - 9 = 0.$

19. $r = \dfrac{6\csc\theta}{2\csc\theta + 3} = \dfrac{6}{2 + 3\sin\theta} \Longrightarrow 2r + 3r\sin\theta = 6 \Longrightarrow \underline{+}2\sqrt{x^2+y^2} =$ $6 - 3y \Longrightarrow 4(x^2+y^2) = 36 - 36y + 9y^2 \Longrightarrow 4x^2 - 5y^2 + 36y - 36 = 0.$

22. The directrix $r = 4\csc\theta$ (or $y = 4$) $\Longrightarrow d = 4$, and the equation is of the form $r = \dfrac{de}{1 + e\sin\theta} = \dfrac{8/5}{1 + (2/5)\sin\theta} = \dfrac{8}{5 + 2\sin\theta}$.

25. The directrix $r\cos\theta = 5$ (or $x = 5$) $\Longrightarrow d = 5$, and the equation is of the form $r = \dfrac{de}{1 + e\cos\theta} = \dfrac{5}{1 + \cos\theta}$.

28. The vertex at $(1, 3\pi/2) \Longrightarrow$ the equation has the form $r = \dfrac{de}{1 \underline{+} e\sin\theta} =$ $\dfrac{\frac{2}{3}d}{1 \underline{+} \frac{2}{3}\sin\theta} = \dfrac{2d}{3 \underline{+} 2\sin\theta}$, since the major axis is along the y-axis. At $\theta = 3\pi/2$, $\sin\theta = -1$ and $r = 1$. Thus we obtain $1 = \dfrac{2d}{3 \overline{+} 2}$ with solution $2d = 5$ or 1. Thus there are 2 possible equations $r = \dfrac{5}{3 - 2\sin\theta}$, $r = \dfrac{1}{3 + 2\sin\theta}$.

31. $y^2 = 4-4x = -4(x-1)$ implies that the graph is a parabola opening to the left with vertex at $(1,0)$. Since $4p = -4$, $p = -1$, and the focus is at the origin, and the directrix is $x = 2$. (Remember, it will be p units to the right of the vertex for such a parabola.) Thus $e = 1$ (as a parabola), $d = 2$ (from the directrix) and the equation is $r = \dfrac{2}{1 + \cos\theta}$. Note that $\cos\theta$ appears, rather than $\sin\theta$, since the directrix is perpendicular to the polar axis. The directrix is $x = r\cos\theta = 2$, or $r = 2\sec\theta$.

34. $5x^2 + 9y^2 = 32x + 64 \Longrightarrow 9x^2 + 9y^2 = 4x^2 + 32x + 64 \Longrightarrow 9(x^2+y^2) =$ $4(x+4)^2 \Longrightarrow \underline{+}3\sqrt{x^2+y^2} = 2(x+4) \Longrightarrow 3r = 2r\cos\theta + 8 \Longrightarrow r(3 - 2\cos\theta) = 8$ $\Longrightarrow r = \dfrac{8}{3 - 2\cos\theta}$. To get d and e, we rewrite as $r = \dfrac{8/3}{1 - (2/3)\cos\theta}$ $\Longrightarrow e = \dfrac{2}{3}$ and $de = \dfrac{8}{3} \Longrightarrow d = 4$. The $\cos\theta$ occurs since the major axis is on the polar axis. The minus sign in the denominator then implies that the directrix is d units to the <u>left</u> of the focus at the pole. Thus the directrix has equation $x = -4 \Longrightarrow r\cos\theta = -4 \Longrightarrow r = -4\sec\theta$.

EXERCISES 13.5, page 636

NOTE: As in Chapters 6 and 10, in addition to the solutions of #1, 4, 7, etc., I have included here and in the next 2 sections the integrals from which the answers to the remaining odd-numbered problems can be obtained.

1. The equation $r = 2\cos\theta$ easily converts to $(x-1)^2 + y^2 = 1$, a circle of radius 1 centered at (1,0). Without integrals the area is π. By the methods of this section, it can be seen from the location of the graph that θ varies from $-\pi/2$ to $\pi/2$. Thus $A = \frac{1}{2}\int_{-\pi/2}^{\pi/2} (2\cos\theta)^2 d\theta = 2\int_{-\pi/2}^{\pi/2} \cos^2\theta\, d\theta$

$$= \int_{-\pi/2}^{\pi/2} (1 + \cos 2\theta)d\theta = \theta + \frac{1}{2}\sin 2\theta\Big]_{-\pi/2}^{\pi/2} = (\frac{\pi}{2} + 0) - (-\frac{\pi}{2} + 0) = \pi.$$

3. $A = 2 \cdot \frac{1}{2}\int_0^{\pi} (1 - \cos\theta)^2 d\theta.$

4. The graph is the cardioid shown. Because of symmetry, we can double the area of the right half

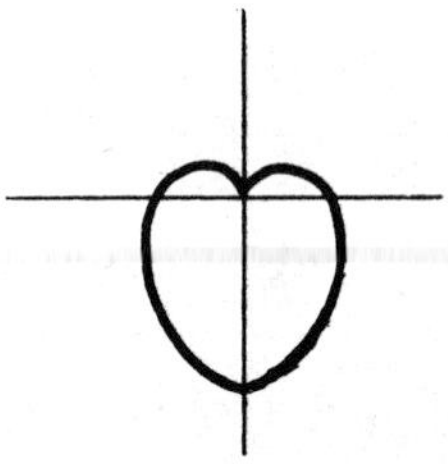

$(-\frac{\pi}{2} \le \theta \le \frac{\pi}{2})$. Thus $A = 2 \cdot \frac{1}{2}\int_{-\pi/2}^{\pi/2} (6-6\sin\theta)^2 d\theta$

$$= 36\int_{-\pi/2}^{\pi/2} (1-2\sin\theta + \sin^2\theta)d\theta =$$

$$36\int_{-\pi/2}^{\pi/2} (1-2\sin\theta + \frac{1-\cos 2\theta}{2})d\theta =$$

$$36\int_{-\pi/2}^{\pi/2} (\frac{3}{2} - 2\sin\theta - \frac{\cos 2\theta}{2})d\theta = 36[\frac{3}{2}\theta + 2\cos\theta - \frac{\sin 2\theta}{4}]_{-\pi/2}^{\pi/2} =$$

$$36[(\frac{3\pi}{4} + 0 - 0) - (-\frac{3\pi}{4} + 0 - 0)] = 36(\frac{3\pi}{2}) = 54\pi.$$

5. $A = 4 \cdot \frac{1}{2}\int_0^{\pi/2} \sin^2 2\theta\, d\theta.$

7. As in #4, we'll double the area of the right half

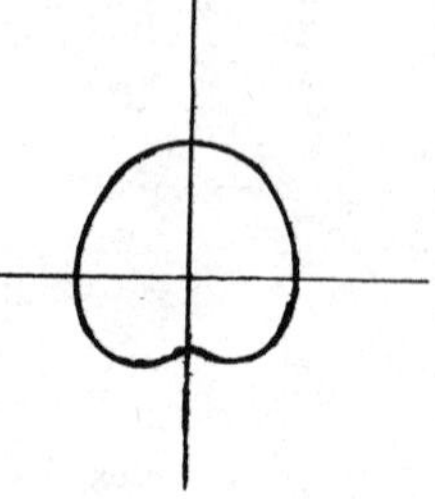

$(-\frac{\pi}{2} \le \theta \le \frac{\pi}{2})$. $A = 2 \cdot \frac{1}{2}\int_{-\pi/2}^{\pi/2} (4 + \sin\theta)^2 d\theta =$

$$\int_{-\pi/2}^{\pi/2} (16+8\sin\theta + \sin^2\theta)d\theta = \int_{-\pi/2}^{\pi/2} (16+8\sin\theta + \frac{1-\cos 2\theta}{2})d\theta$$

$$= \int_{-\pi/2}^{\pi/2} (\frac{33}{2} + 8\sin\theta - \frac{\cos 2\theta}{2})d\theta =$$

$$[\frac{33}{2}\theta - 8\cos\theta - \frac{\sin 2\theta}{4}]_{-\pi/2}^{\pi/2} = \frac{33}{2}\pi.$$

9. $A = \frac{1}{2}\int_0^{\pi/2} e^{2\theta}\, d\theta.$

10. $A = \frac{1}{2}\int_0^{\pi} (2\theta)^2 d\theta = \frac{2}{3}\theta^3\Big]_0^{\pi} = \frac{2}{3}\pi^3.$

11. $A = \frac{1}{2}\int_{-\pi/4}^{\pi/4} 4\cos 2\theta\, d\theta.$

13. We'll take the loop framed by $\theta = \pm\frac{\pi}{10}$.

$A = \frac{1}{2}\int_{-\pi/10}^{\pi/10} 3^2\cos^2 5\theta\, d\theta = \frac{9}{4}\int_{-\pi/10}^{\pi/10} (1 + \cos 10\theta)d\theta$

$= \frac{9}{4}[\theta + \frac{\sin 10\theta}{10}]_{-\pi/10}^{\pi/10} = \frac{9\pi}{20}.$

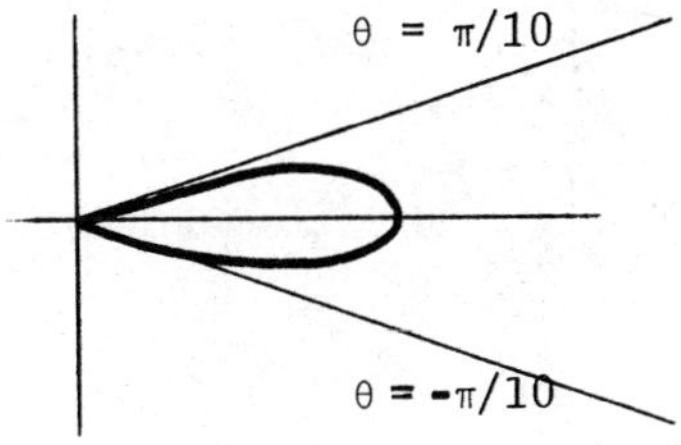

15. $A = 2 \cdot \frac{1}{2}\int_{\pi/3}^{\pi} (3^2 - (2 + 2\cos\theta)^2)\, d\theta.$

16. The values of θ at the points of intersection are found by setting $2 = 4\cos\theta \Longrightarrow \cos\theta = 1/2 \Longrightarrow \theta = \pm\pi/3$.

$A = \frac{1}{2}\int_{-\pi/3}^{\pi/3} [(4\cos\theta)^2 - 2^2]d\theta$

$= 2\int_{-\pi/3}^{\pi/3} (4\cos^2\theta - 1)d\theta$

$= 2\int_{-\pi/3}^{\pi/3} (1 + 2\cos 2\theta)d\theta$

$= 2[\theta + \sin 2\theta]_{-\pi/3}^{\pi/3} = \frac{4\pi}{3} + 2\sqrt{3}.$

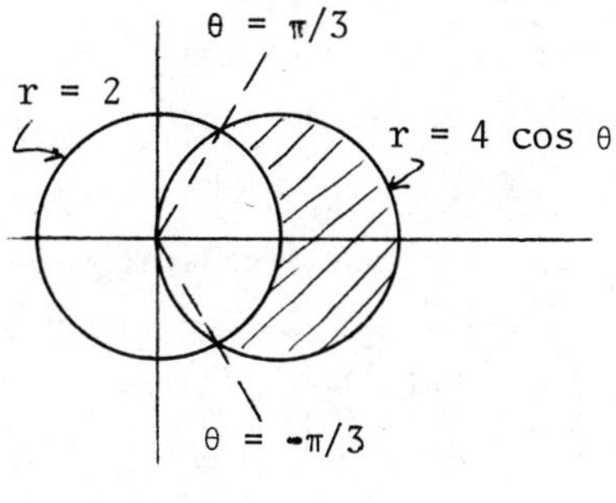

17. $A = 2 \cdot \frac{1}{2}\int_{\pi/12}^{5\pi/12} (8\sin 2\theta - 2^2)d\theta.$

19. The value of θ at the point of intersection in the 1st quadrant is the solution of $\sin\theta = \sqrt{3}\cos\theta$ yielding $\tan\theta = \sqrt{3}$ or $\theta = \pi/3$. For $0 \le \theta \le \pi/3$, the boundary curve is $r = \sin\theta$ whereas for $\frac{\pi}{3} \le \theta \le \frac{\pi}{2}$, The boundary is $r = \sqrt{3}\cos\theta$.

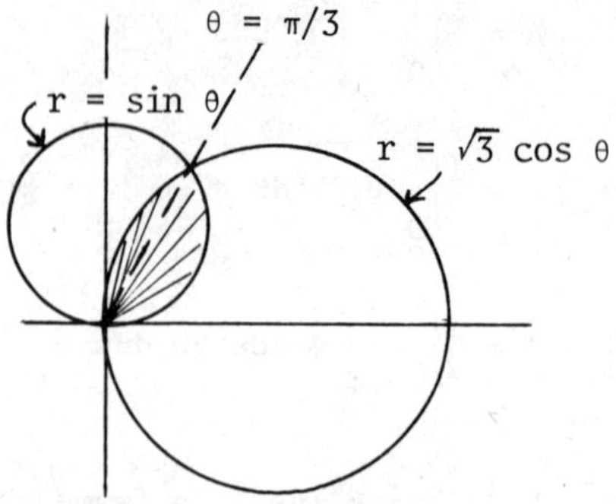

Thus $A = \frac{1}{2}\int_0^{\pi/3} \sin^2\theta\, d\theta + \frac{1}{2}\int_{\pi/3}^{\pi/2} 3\cos^2\theta\, d\theta = \frac{1}{4}\int_0^{\pi/3} (1 - \cos 2\theta)d\theta + \frac{3}{4}\int_{\pi/3}^{\pi/2} (1 + \cos 2\theta)d\theta = \frac{1}{4}[\theta - \frac{\sin 2\theta}{2}]_0^{\pi/3}$

$+ \frac{3}{4}[\theta + \frac{\sin 2\theta}{2}]_{\pi/3}^{\pi/2} = \frac{1}{4}[\frac{\pi}{3} - \frac{\sqrt{3}}{4}] + \frac{3}{4}[\frac{\pi}{6} - \frac{\sqrt{3}}{4}] = \frac{5\pi}{24} - \frac{\sqrt{3}}{4}$.

21. $A = 2 \cdot \frac{1}{2}\int_0^{\arcsin(1/4)} (5\sin\theta)^2 d\theta + 2 \cdot \frac{1}{2}\int_{\arcsin(1/4)}^{\pi/2} (1 + \sin\theta)^2 d\theta$.

22. We'll calculate the area in the 1st quadrant and multiply by 4. The curves intersect in the 1st quadrant when $4\cos 2\theta = 1 \Longrightarrow 2\theta = \cos^{-1}(1/4)$. Let $\theta_o = \frac{1}{2}\cos^{-1}(1/4) \approx 37^o 45'$. For $0 \le \theta \le \theta_o$, $r = 1$ is the bounding curve, and for $\theta_o \le \theta \le \pi/4$, $r^2 = 4\cos 2\theta$ is the bounding curve. Thus

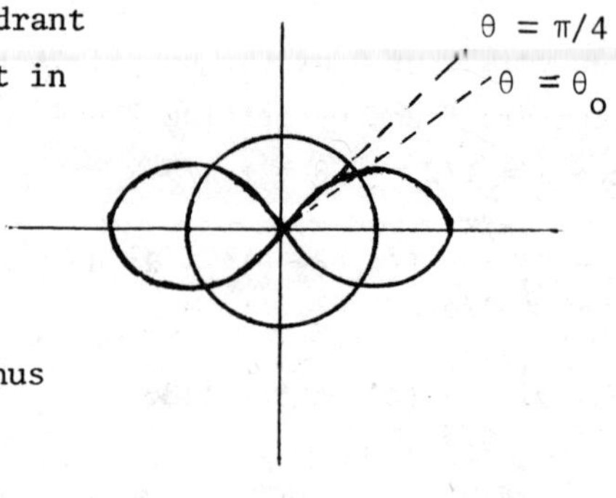

$A = 4[\frac{1}{2}\int_0^{\theta_o} 1\, d\theta + \frac{1}{2}\int_{\theta_o}^{\pi/4} 4\cos 2\theta\, d\theta] =$

$2\theta_o + 4\sin 2\theta \Big]_{\theta_o}^{\pi/4} = \cos^{-1}\frac{1}{4} + 4 - 4\sin(\cos^{-1}\frac{1}{4}) \approx 1.318 + 4 - 3.873 = 1.445.$

23. $A = \frac{1}{2}\int_{\pi/6}^{\pi/3} 4\sec^2\theta\, d\theta.$

25. The only reasonable way to solve this problem is to rewrite the equations in rectangular form and then use the methods of Section 6.1. Otherwise, the attempt to evaluate $A = 8\int_{\pi/4}^{5\pi/4} \frac{1}{(1 - \cos\theta)^2}\, d\theta$ by the substitution $z = \tan\theta/2$ leads to a difficult improper integral. If one uses trigonometric identities to rewrite $\frac{1}{(1-\cos\theta)^2} = \frac{(1+\cos\theta)^2}{\sin^4\theta}$, the new form of the integrand is discontinuous at $\theta = \pi$ and another improper integral results.

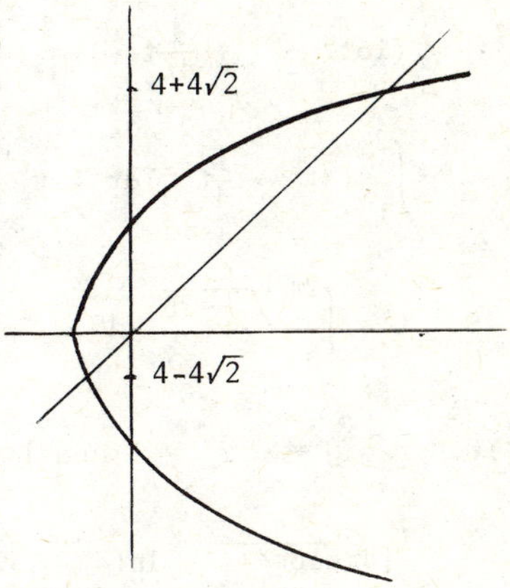

Numerical integration by Simpson's Rule is a viable alternative. Now,

$r = \frac{4}{1 - \cos\theta} \Longrightarrow r - r\cos\theta = 4 \Longrightarrow r = x+4 \Longrightarrow x^2+y^2 = x^2 + 8x + 16 \Longrightarrow$

$x = (y^2-16)/8$. $\theta = \pi/4$ is the line $y = x$. The points of intersection are at solutions of $y = (y^2 - 16)/8$ or $y = 4 \pm 4\sqrt{2}$. Thus

$$A = \int_{4-4\sqrt{2}}^{4+4\sqrt{2}} (y - \frac{y^2-16}{8})\,dy = \frac{y^2}{2} - \frac{y^3}{24} + 2y\Big|_{4-4\sqrt{2}}^{4+4\sqrt{2}} = \frac{64\sqrt{2}}{3}.$$

EXERCISES 13.6, page 640

1. $x = 5t^2 \Longrightarrow x' = 10t$, $y = 2t^3 \Longrightarrow y' = 6t^2$, $0 \le t \le 1$.

$$L = \int_0^1 \sqrt{(10t)^2 + (6t^2)^2}\, dt = \int_0^1 2t\sqrt{25 + 9t^2}\, dt.$$

Let $u = 25 + 9t^2$, $du = 18t\, dt$, $2t\, dt = du/9$. Then

$$L = \frac{1}{9}\int_{25}^{34} u^{1/2} du = \frac{1}{9}(\frac{2}{3})u^{3/2}\Big]_{25}^{34} = \frac{2}{27}(34^{3/2} - 25^{3/2}).$$

3. $$L = \int_0^{\pi/2} \sqrt{e^{2t}(\cos t - \sin t)^2 + e^{2t}(\cos t + \sin t)^2}\, dt.$$

4. $x' = 6t$, $y' = 4t$, $0 \le t \le 2 \Longrightarrow L = \int_0^2 \sqrt{36t^2 + 16t^2}\, dt = \sqrt{52}\int_0^2 \sqrt{t^2}\, dt.$

Recall that $\sqrt{t^2} = |t|$. Since $0 \le t \le 2$ here, $\sqrt{t^2} = t$. Thus

$$L = \sqrt{52}\int_0^2 t\, dt = \sqrt{52}\,\frac{t^2}{2}\Big|_0^2 = 2\sqrt{52} = 2\sqrt{4(13)} = 4\sqrt{13}.$$

5. $$L = \int_0^{\pi/3} \sqrt{\tan^2 t + 1}\, dt.$$

7. $x' = 2$, $y' = 4t^3 - \frac{1}{4}t^{-3} \Longrightarrow x'^2 + y'^2 = 4 + (16t^6 - 2 + \frac{1}{16}t^{-6}) =$

$(16t^6 + 2 + \frac{1}{16}t^{-6}) = (4t^3 + (\frac{1}{4})t^{-3})^2$. Thus $L = \int_1^2 \sqrt{x'^2 + y'^2}\,dt =$

$\int_1^2 (4t^3 + \frac{1}{4}t^{-3})dt = t^4 - \frac{t^{-2}}{8}\Big|_1^2 = (16 - \frac{1}{32}) - (1 - \frac{1}{8}) = 15\frac{3}{32} = \frac{483}{32}$.

9. $L = \int_0^{2\pi} \sqrt{2e^{-2\theta}}\, d\theta.$

10. $r = \theta \Longrightarrow r' = 1$ and $L = \int_0^{4\pi} \sqrt{\theta^2+1}\, d\theta = \frac{1}{2}[\theta\sqrt{\theta^2+1} + \ln(\theta + \sqrt{\theta^2+1})]_0^{4\pi} =$

$\frac{1}{2}[4\pi\sqrt{16\pi^2+1} + \ln(4\pi + \sqrt{16\pi^2+1})]$. Formula 21, or a trig substitution, provides the antiderivative.

11. $L = \int_0^{\pi} \sqrt{\cos^4 \frac{\theta}{2} + \cos^2 \frac{\theta}{2} \sin^2 \frac{\theta}{2}}\, d\theta$.

13. The distance that the particle travels is, of course, the length of the curve upon which it moves. $x = 4t+3 \Longrightarrow x' = 4$, $y = 2t^2 \Longrightarrow y' = 4t$, $0 \le t \le 5$.

$L = \int_0^5 \sqrt{16 + 16t^2}\, dt = 4\int_0^5 \sqrt{1 + t^2}\, dt = 2[t\sqrt{1+t^2} + \ln(t + \sqrt{1+t^2})]_0^5 =$

$2[5\sqrt{26} + \ln(5 + \sqrt{26})]$.

14. $x = \cos 2t \Longrightarrow x' = -2 \sin 2t$, $y = \sin^2 t \Longrightarrow y' = 2 \sin t \cos t = \sin 2t$,

$0 \le t \le \pi$. $L = \int_0^{\pi} \sqrt{4 \sin^2 2t + \sin^2 2t}\, dt = \sqrt{5}\int_0^{\pi} |\sin 2t|\,dt =$

$\sqrt{5}\int_0^{\pi/2} \sin 2t\, dt - \sqrt{5}\int_{\pi/2}^{\pi} \sin 2t\, dt = -\frac{\sqrt{5}}{2}\cos 2t]_0^{\pi/2} + \frac{\sqrt{5}}{2}\cos 2t]_{\pi/2}^{\pi} = 2\sqrt{5}.$

15. $L = \int_0^{\pi/2} \sqrt{t^2 \sin^2 t + t^2 \cos^2 t}\, dt.$

16. Very similar to #13 above. Answer: $L = 4\sqrt{17} + \ln(4 + \sqrt{17})$.

19. $x = 1/t \Longrightarrow x' = -1/t^2$, $y = \ln t \Longrightarrow y' = 1/t$, $1 \le t \le 2$.

$L = \int_1^2 \sqrt{\frac{1}{t^4} + \frac{1}{t^2}}\, dt = \int_1^2 \frac{\sqrt{1+t^2}}{t^2}\, dt.$ This integral can be evaluated by Formula 24 or by a trig substitution followed by a variable change and a partial fraction decomposition. In any event, $L = [-\frac{\sqrt{1+t^2}}{t} + \ln(\sqrt{1+t^2} + t)]_1^2 =$

$-\frac{\sqrt{5}}{2} + \ln(\sqrt{5} + 2) + \sqrt{2} - \ln(\sqrt{2} + 1).$

EXERCISES 13.7, page 645

1. $x = t^2 \Longrightarrow x' = 2t$, $y = 2t \Longrightarrow y' = 2$, $0 \le t \le 4$. Thus $S = 2\pi\int_0^4 y\sqrt{x'^2+y'^2}\,dt$

$= 2\pi\int_0^4 2t\sqrt{(2t)^2+2^2}\,dt = 4\pi\int_0^4 (2t)\sqrt{t^2+1}\,dt = \frac{8\pi}{3}(t^2+1)^{3/2}\Big]_0^4 = \frac{8\pi}{3}(17^{3/2} - 1).$

3. $S = 2\pi\int_0^1 (t - \frac{1}{3}t^3)\sqrt{(t^2+1)^2}\,dt.$

4. $x' = 8t$, $y = 3-2t$, $y' = -2 \Longrightarrow S = 2\pi\int_{-2}^{0} (3-2t)\sqrt{64t^2+4}\,dt =$

$6\pi\int_{-2}^{0} \sqrt{64t^2+4}\,dt - 4\pi\int_{-2}^{0} t\sqrt{64t^2+4}\,dt = I_1 - I_2$. For I_1, let $u = 8t$ $(dt = du/8)$.
When $t = -2,0$, $u = -16,0$, and, using Formula 21, $I_1 = \frac{6\pi}{8}\int_{-16}^{0} \sqrt{u^2+4}\,du =$

$\frac{3\pi}{4}[\frac{u}{2}\sqrt{u^2+4} + \frac{4}{2}\ln|u + \sqrt{u^2+4}|]_{-16}^{0} = \frac{3\pi}{4}[(0 + 2\ln\sqrt{4}) - (-8\sqrt{260} + 2\ln|-16 + \sqrt{260}|)].$

Writing $\sqrt{260} = \sqrt{4(65)} = 2\sqrt{65}$, and $\ln|-16 + \sqrt{260}| = \ln|2(\sqrt{65}-8)| = \ln 2 +$ $\ln(\sqrt{65}-8)$, we obtain: $I_1 = \frac{3\pi}{2}(8\sqrt{65} - \ln(\sqrt{65}-8))$. For I_2, let $u = 64t^2+4$,

$t\,dt = \frac{du}{128}$. When $t = -2,0$, $u = 260,4$ and $I_2 = \frac{4\pi}{128}\int_{260}^{4} \sqrt{u}\,du =$

$\frac{\pi}{32}\cdot\frac{2}{3}u^{3/2}\Big]_{260}^{4} = \frac{\pi}{48}(4^{3/2}-260^{3/2}) = \frac{\pi}{6}(1-65^{3/2}) = \frac{\pi}{6} - \frac{\pi}{6}\cdot 65\sqrt{65}$. Combining results: $S = I_1-I_2 = 12\pi\sqrt{65} - \frac{3\pi}{2}\ln(\sqrt{65}-8) - \frac{\pi}{6} + \frac{\pi}{6}65\sqrt{65} =$

$\frac{(137\sqrt{65}-1)\pi}{6} - \frac{3\pi}{2}\ln(\sqrt{65}-8) \approx 590.89.$

5. $S = 2\pi\int_1^2 x^3\sqrt{1 + 9x^4}\,dx.$

7. $x = a(t - \sin t) \Longrightarrow x' = a(1 - \cos t)$, $y = a(1 - \cos t) \Longrightarrow y' = a\sin t$, $0 \le t \le 2\pi$.

$$S = 2\pi\int_0^{2\pi} a(1 - \cos t)\sqrt{a^2(1 - \cos t)^2 + a^2\sin^2 t}\,dt$$

$$= 2\pi a^2\int_0^{2\pi} (1 - \cos t)\sqrt{1 - 2\cos t + \underbrace{\cos^2 t + \sin^2 t}_{=1}}\,dt$$

$= 2\sqrt{2}\pi a^2\int_0^{2\pi} (1 - \cos t)^{3/2}\,dt$. Now, using $\cos t = 1 - 2\sin^2\frac{t}{2}$, the integrand is $2^{3/2}\sin^3\frac{t}{2}$. The integral is then done as in Sec. 10.2

writing $\sin^3 = (1 - \cos^2)\sin$. Ultimately we obtain

$$S = 16\pi a^2\left[-\cos\frac{t}{2} + \frac{1}{3}\cos^3\frac{t}{2}\right]_0^{2\pi} = \frac{64\pi a^2}{3}.$$

9. $S = 2\pi\int_1^4 4t^{1/2}\,(t + t^{-2})dt.$

10. $x' = 3,\ y' = 1 \Longrightarrow S = 2\pi\int_0^5 3t\sqrt{3^2+1^2}\,dt = 6\sqrt{10}\,\pi\,\frac{t^2}{2}\Big]_0^5 = 75\sqrt{10}\,\pi.$

11. $S = 2\pi\int_0^{\pi/2} e^t \sin t\sqrt{2e^{2t}}\,dt.$

13. By (13.26) with $y' = x/8$, $S = 2\pi\int_0^8 x\sqrt{1 + \frac{x^2}{64}}\,dx = 64\pi\left(\frac{2}{3}\right)\left(1 + \frac{x^2}{64}\right)^{3/2}\Big]_0^8$

$= \frac{128\pi}{3}(2^{3/2} - 1).$

15. $S = 2\pi\int_1^2 x\sqrt{1 + 1/x^2}\,dx.$

16. $y' = \sinh x$ and $S = 2\pi\int_1^2 x\sqrt{1 + \sinh^2 x}\,dx = 2\pi\int_1^2 x\cosh x\,dx$. Now, integrate by parts with $u = x$, $dv = \cosh x\,dx$ so that $du = dx$, $v = \sinh x$, and

$S = 2\pi\left(x\sinh x\Big]_1^2 - \int_1^2 \sinh x\,dx\right) = 2\pi[x\sinh x - \cosh x]_1^2 =$

$2\pi[(2\sinh 2 - \cosh 2) - (\sinh 1 - \cosh 1)].$

17. $S = 2\pi\int_0^{\pi} (2 + 2\cos\theta)\sin\theta\sqrt{8(1 + \cos\theta)}\,d\theta.$

19. $r = 2a\sin\theta = f(\theta) \Longrightarrow f'(\theta) = 2a\cos\theta$ and, assuming $a > 0$,

$S = \int_0^{\pi} 2\pi(2a\sin\theta)\sin\theta\sqrt{4a^2(\sin^2\theta + \cos^2\theta)}\,d\theta = 8\pi a^2\int_0^{\pi}\sin^2\theta\,d\theta =$

$4\pi a^2\int_0^{\pi}(1 - \cos 2\theta)d\theta = 4\pi^2 a^2.$

21. $S = 2\pi\int_{\arctan(4/3)}^{\pi/2} 5\cos t(5)\,dt$ (using parametric equations $x = 5\cos t$, $y = 5\sin t$ for the circle).

22. Using parametric equations $x = 5\cos t$, $y = 5\sin t$ for the circle, the arc from (3,4) to (0,5) results when t ranges from $\tan^{-1}\left(\frac{4}{3}\right)$ to $\frac{\pi}{2}$. This is $\frac{1}{2}$ of the entire arc, so we can use these limits and multiply the result by 2.

Since $x'^2 + y'^2 = 25$ we obtain $\frac{S}{2} = 2\pi\int_{\tan^{-1}\frac{4}{3}}^{\pi/2} (5\sin t)(5)dt =$

$50\pi[-\cos t]_{\tan^{-1}\frac{4}{3}}^{\pi/2}$. By elementary trig, $\cos(\tan^{-1}\frac{4}{3}) = \frac{3}{5}$, and thus

$S = 100\pi(0 + \frac{3}{5}) = 60\pi$.

25. We use the parametric equations $x = a\cos t$, $y = a\sin t$, $0 \le t \le 2\pi$ for the circle. The derivation of the formula for this problem is similar to the derivation of (13.24), but instead of the frustrum of a cone, we have to estimate the slant surface area of the cylindrical shell traced out by the trapezoid shown. The slant distance is still $\approx \sqrt{f'^2 + g'^2}\,\Delta t_i$

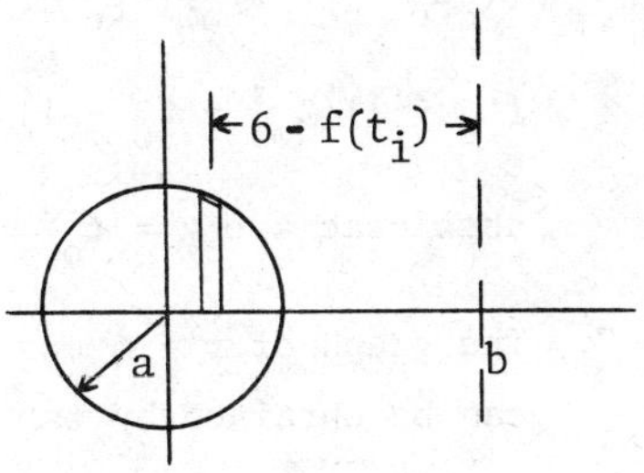

but the radius is now $\approx b - f(t_i)$. Combining this as in the discussion of

(13.24) we obtain $S = 2\pi\int_0^{2\pi} (b - f(t))\sqrt{f'(t)^2 + g'(t)^2}\,dt =$

$2\pi\int_0^{2\pi} (b - a\cos t)\sqrt{a^2\sin^2 t + a^2\cos^2 t}\,dt = 2\pi a[bt - a\sin t]_0^{2\pi} =$

$2\pi a(2\pi b) = 4\pi^2 ab$.

EXERCISES 13.8 (Review), page 646

1. $x = \frac{1}{t} + 1 \Rightarrow t = \frac{1}{x-1}$. Thus $y = \frac{2}{t} - t =$

$2(x-1) - \frac{1}{x-1}$ is the rectangular equation. The value $t = 4$ yields the point $(\frac{5}{4}, -\frac{7}{2})$. As t decreases to 0 the point (x,y) moves upward on the graph.

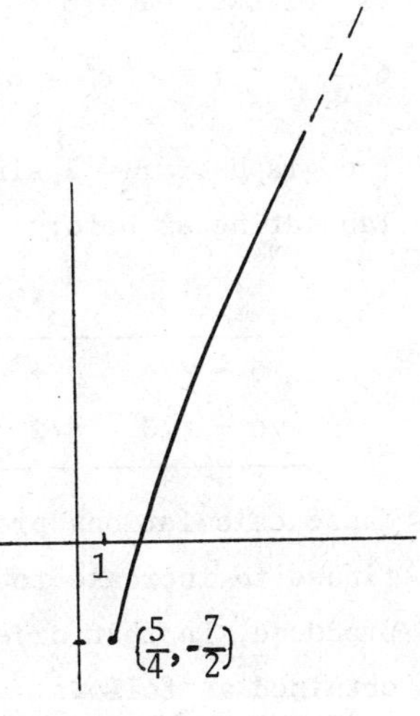

$$\left.\frac{dx}{dt}\right|_{t=1} = -\left.\frac{1}{t^2}\right|_{t=1} = -1;$$

$$\left.\frac{dy}{dt}\right|_{t=1} = \left.-\frac{2}{t^2} - 1\right|_{t=1} = -3.$$

Thus $\left.\frac{dy}{dx}\right|_{t=1} = \frac{-3}{-1} = 3$.

4. $x' = 2t$, $y' = 6t^2+4 \Longrightarrow \frac{dy}{dt} = \frac{6t^2+4}{2t} = \frac{3t^2+2}{t}$. Let the point of tangency correspond to $t = t_o$. Then the point is $(t_o^2, 2t_o^3 + 4t_o - 1)$ and the equation of the tangent line is $y - (2t_o^3 + 4t_o - 1) = (\frac{3t_o^2 + 2}{t_o})(x - t_o^2)$. Since this line is to pass through the origin, $x = 0$ and $y = 0$ must satisfy this equation. Substituting, we get the equation for t_o: $-(2t_o^3 + 4t_o - 1) = [(3t_o^2 + 2)/t_o](-t_o^2) = (-3t_o^3 - 2t_o)$ or $t_o^3 - 2t_o + 1 = (t_o - 1)(t_o^2 + t_o - 1) = 0$, with solutions $t_o = 1, \frac{-1+\sqrt{5}}{2}$. Thus the desired absciccas are $x = t_o^2 = 1, \frac{3+\sqrt{5}}{2}$.

7. The graph of $r = 6 \cos 2\theta$, a 4-leaved rose, can be obtained by tabulating as in Sec. 13.3.

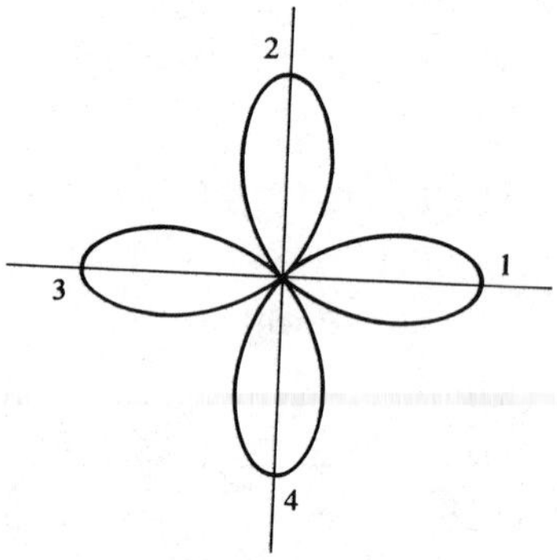

θ	2θ	$r = 6 \cos 2\theta$
$-\frac{\pi}{4} \to 0$	$-\frac{\pi}{2} \to 0$	$0 \to 6$
$0 \to \frac{\pi}{4}$	$0 \to \frac{\pi}{2}$	$6 \to 0$

These calculations produce leaf 1 in the sketch. As θ continues increasing to π, leaves 4, 3, and 2 are next produced, in that order. The xy equation is obtained as follows: $r = 6 \cos 2\theta = 6(\cos^2\theta - \sin^2\theta) = 6(\frac{x^2}{r^2} - \frac{y^2}{r^2}) \Longrightarrow r^3 = 6(x^2 - y^2) \Longrightarrow (x^2 + y^2)^{3/2} = 6(x^2 - y^2)$.

10. The graph of $r = 2 \sin 3\theta$ is a 3-leaved rose. Tabulating as before:

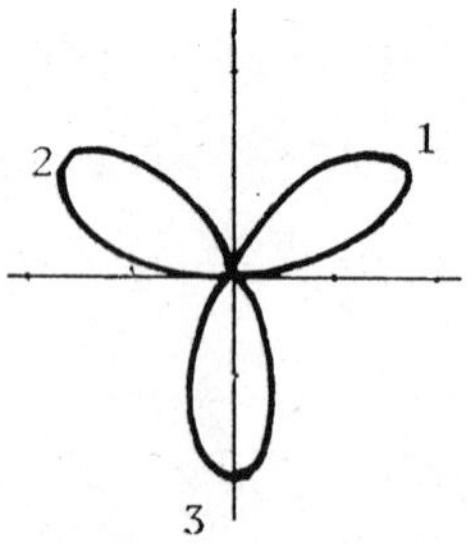

θ	3θ	$r = 2 \sin 3\theta$
$0 \to \pi/6$	$0 \to \pi/2$	$0 \to 2$
$\pi/6 \to \pi/3$	$\pi/2 \to \pi$	$2 \to 0$

These calculations produce leaf 1. As θ continues to increase to π, leaves 3 and 2 are produced, in that order. The xy equation is obtained as follows:

$r = 2 \sin 3\theta = 2(3 \sin \theta - 4 \sin^3\theta) = 6\frac{y}{r} - 8\frac{y^3}{r^3} \Longrightarrow r^4 = 6yr^2 - 8y^3 \Longrightarrow$

$(x^2 + y^2)^2 = 6y(x^2 + y^2) - 8y^3 = 6x^2y - 2y^3$.

13. $r^2 = \sec 2\theta = \dfrac{1}{\cos 2\theta} = \dfrac{1}{\cos^2\theta - \sin^2\theta} \Rightarrow r^2\cos^2\theta - r^2\sin^2\theta = 1 \Rightarrow x^2-y^2 = 1,$ a familiar hyperbola.

16. That the graph is a parabola opening to the left with vertex (3,0) can be seen by writing the equation in the form $r = \dfrac{6}{1 + \cos\theta}$. The xy equation is obtained from the given equation: $r = 6 - r\cos\theta = 6 - x \Rightarrow \pm\sqrt{x^2 + y^2} = 6 - x \Rightarrow x^2 + y^2 = 36 - 12x + x^2 \Rightarrow y^2 = -12(x-3)$.

19. $r = \dfrac{8}{3 + \cos\theta} \Rightarrow 3r + r\cos\theta = 8 \Rightarrow 3r = 8 - x \Rightarrow 9(x^2 + y^2) = 64 - 16x + x^2 \Rightarrow 8x^2 + 16x + 9y^2 = 64$, an ellipse, as was obvious from the given equation ($e = 1/3$).

22. Directrix $r = 6\sec\theta \Rightarrow d = 6$ and the polar equation is the "$+e\cos\theta$" form. So, with $e = 2$ we get $r = \dfrac{de}{1 + e\cos\theta} = \dfrac{12}{1 + 2\cos\theta}$.

25. The distance traveled is the length of the path of motion. $x' = 2\cos t$, $y' = 2\sin t\cos t \Rightarrow \sqrt{x'^2 + y'^2} = \sqrt{4\cos^2 t(1 + \sin^2 t)} = 2\cos t\sqrt{1 + \sin^2 t}$ since $0 \le t \le \pi/2$ and $\cos t \ge 0$ there. $L = \displaystyle\int_0^{\pi/2} 2\cos t\sqrt{1 + \sin^2 t}\,dt = 2\int_0^1 \sqrt{1 + u^2}\,du$, where $u = \sin t$. By formula 21, $L = [u\sqrt{1 + u^2} + \ln(u + \sqrt{1 + u^2})]_0^1 = \sqrt{2} + \ln(1 + \sqrt{2})$.

28. $x' = 4t,\ y' = 4 \Rightarrow \sqrt{x'^2 + y'^2} = \sqrt{16t^2 + 16} = 4\sqrt{t^2 + 1} \Rightarrow S = 8\pi\displaystyle\int_0^1 (2t^2 + 1)\sqrt{t^2+1}\,dt = 16\pi\int_0^1 t^2\sqrt{t^2 + 1}\,dt + 8\pi\int_0^1 \sqrt{t^2 + 1}\,dt$. Using Formulas 21 and 22, we obtain $S = 8\pi[\frac{t}{2}\sqrt{(t^2+1)^3} + \frac{1}{4}t\sqrt{t^2+1} + \frac{1}{4}\ln(t + \sqrt{t^2+1})]_0^1 = 8\pi[\frac{\sqrt{8}}{2} + \frac{\sqrt{2}}{4} + \frac{1}{4}\ln(1 + \sqrt{2})] = 2\pi[5\sqrt{2} + \ln(1 + \sqrt{2})]$.

CHAPTER 14

VECTORS AND SOLID ANALYTIC GEOMETRY

NOTE: Throughout Chapters 14-18, vectors will be denoted by underlining as $\underline{a}$, $\underline{b}$, etc. A directed line segment from point P to point Q will be denoted as $\overrightarrow{PQ}$.

EXERCISES 14.1, page 659

1. (ii) This is immediate from the definition of addition of vectors and the associativity of addition for real numbers which gives us $a_i + (b_i + c_i) = (a_i + b_i) + c_i$ for i = 1 and 2.

   (iii) $\underline{a} + \underline{0} = \langle a_1,a_2\rangle + \langle 0,0\rangle = \langle a_1+0,a_2+0\rangle = \langle a_1,a_2\rangle = \underline{a}$.

   (iv) $\underline{a} + (-\underline{a}) = \langle a_1,a_2\rangle + \langle -a_1,-a_2\rangle = \langle a_1+(-a_1),a_2+(-a_2)\rangle = \langle 0,0\rangle = \underline{0}$.

4. $(-c)\underline{a} = (-c)\langle a_1,a_2\rangle = \langle (-c)a_1,(-c)a_2\rangle = \langle -ca_1,-ca_2\rangle = -\langle ca_1,ca_2\rangle = -(c\underline{a})$.

7. $\underline{a} + \underline{b} = \underline{0} \iff \langle a_1+b_1,a_2+b_2\rangle = \langle 0,0\rangle \iff a_1 + b_1 = 0$ and $a_2 + b_2 = 0$ $\iff b_1 = -a_1,\ b_2 = -a_2 \iff \underline{b} = \langle -a_1,-a_2\rangle = -\underline{a}$.

10. $\underline{a} \neq \underline{0} \Rightarrow a_1$ and a_2 are not both 0. Suppose $a_1 \neq 0$. Then $c\underline{a} = \underline{0} \iff \langle ca_1,ca_2\rangle = \underline{0} \Rightarrow ca_1 = 0 \Rightarrow c = 0$. Similarly, if $a_2 \neq 0$, then c = 0.

13. With $\underline{a} = \langle 7, 2\rangle = \langle -7,2\rangle$, $\underline{b} = 4\langle -2,1\rangle = \langle -8,4\rangle$ we get. $\underline{a} + \underline{b} = \langle -7+(-8),2+4\rangle = \langle -15,6\rangle$, $\underline{a} - \underline{b} = \langle -7-(-8),2-4\rangle = \langle 1,-2\rangle$, $4\underline{a} + 5\underline{b} = \langle -28,8\rangle + \langle -40,20\rangle = \langle -68,28\rangle$, $4\underline{a} - 5\underline{b} = \langle -28,8\rangle - \langle -40,20\rangle = \langle -28+40,8-20\rangle = \langle 12,-12\rangle$.

16. Here $\underline{a} = \underline{b} = -3\underline{i} + \underline{j}$. Thus $\underline{a} + \underline{b} = 2\underline{a} = -6\underline{i} + 2\underline{j}$, $\underline{a} - \underline{b} = \underline{0}$, $4\underline{a} + 5\underline{b} = 9\underline{a} = -27\underline{i} + 9\underline{j}$, $4\underline{a} - 5\underline{b} = -\underline{a} = 3\underline{i} - \underline{j}$.

19. With $\underline{a} = 2\underline{j}$, $\underline{b} = -3\underline{i}$, we get: $\underline{a} + \underline{b} = 2\underline{j} + (-3\underline{i}) = -3\underline{i} + 2\underline{j}$, $\underline{a} - \underline{b} = 2\underline{j} - (-3\underline{i}) = 3\underline{i} + 2\underline{j}$, $4\underline{a} \pm 5\underline{b} = 8\underline{j} \mp 15\underline{i} = \mp 15\underline{i} + 8\underline{j}$.

22. As in Example 2, $\overrightarrow{PQ} = \langle -2-7,4-(-3)\rangle = \langle -9,7\rangle = -9\underline{i} + 7\underline{j}$.

25. $\overrightarrow{PQ} = \langle 6-(-3),-4-(-1)\rangle = \langle 9,-3\rangle = 9\underline{i} - 3\underline{j}$.

28. $\underline{a} + \underline{b} = -4\underline{i} - \underline{j}$, $\underline{a} - \underline{b} = -6\underline{i} + 5\underline{j}$, $2\underline{a} = -10\underline{i} + 4\underline{j}$, $-3\underline{b} = -3\underline{i} + 9\underline{j}$. See sketches below.

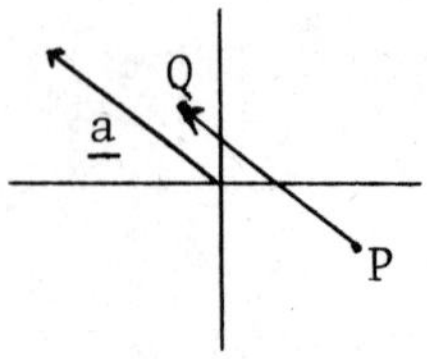

Ex. 22

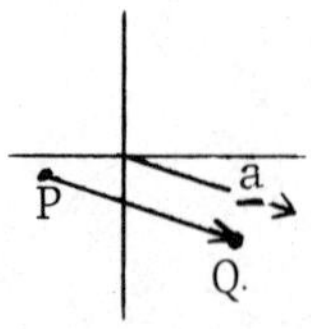

Ex. 25

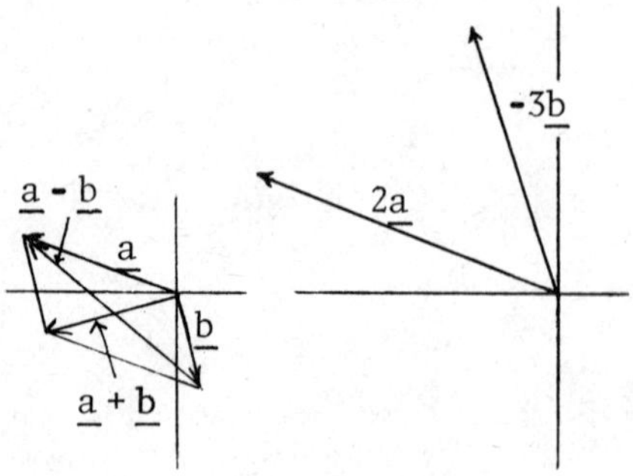

Exercise 28

31. $|\underline{a}| = \sqrt{3^2+(-3)^2} = \sqrt{18}$. 34. $|\underline{a}| = \sqrt{0+10^2} = 10$. 37. $|\underline{a}| = \sqrt{0^2+18^2} = 18$.

40. $|\underline{a}| = \sqrt{5^2+3^2} = \sqrt{34}$. (a) $\underline{u} = \frac{1}{|\underline{a}|}\underline{a} = \frac{1}{\sqrt{34}}(5\underline{i}-3\underline{j}) = \frac{5}{\sqrt{34}}\underline{i} - \frac{3}{\sqrt{34}}\underline{j}$ is a unit vector having the same direction as $\underline{a}$. (b) $-\underline{u} = -\frac{5}{\sqrt{34}}\underline{i} + \frac{3}{\sqrt{34}}\underline{j}$ is a unit vector having the direction opposite to $\underline{a}$.

43. (a) By (14.6i) and (14.7), $2\underline{a} = \langle -12,6\rangle$ has the same direction as $\underline{a}$ and twice the magnitude of $\underline{a}$ (since $|2\underline{a}| = 2|\underline{a}|$). (b) Here, $\frac{1}{2}\underline{a} = \langle -3,1.5\rangle$ is the desired vector.

46. If $\underline{u} = \frac{1}{|\underline{a}|}\underline{a}$, then $-4\underline{u}$ is the desired vector. $|\underline{a}| = \sqrt{4+25} = \sqrt{29} \Rightarrow$ $\underline{u} = \left\langle \frac{2}{\sqrt{29}}, \frac{-5}{\sqrt{29}} \right\rangle \Rightarrow -4\underline{u} = \left\langle \frac{-8}{\sqrt{29}}, \frac{20}{\sqrt{29}} \right\rangle$.

EXERCISES 14.2, page 665

1. (a) $d(A,B) = \sqrt{(2-4)^2 + (4-(-2))^2 + (-5-3)^2} = \sqrt{4+36+64} = \sqrt{104}$.
(b) Midpoint $= (\frac{2+4}{2}, \frac{4-2}{2}, \frac{-5+3}{2}) = (3,1,-1)$.

4. (a) $d(A,B) = \sqrt{(1-0)^2 + (1-5)^2 + (0-(-4))^2} = \sqrt{1+16+16} = \sqrt{33}$.
(b) Midpoint $= (\frac{1+0}{2}, \frac{1+5}{2}, \frac{0-4}{2}) = (1/2,3,-2)$.

7. The faces of the box shown have the following equations: front, $x = 2$; back, $x = -4$; left, $y = 2$; right, $y = 5$; top, $z = 1$; bottom, $z = -3$. (The 1st, 4th, and 6th come from the point A, the others from B.) These now yield the other six points as given.

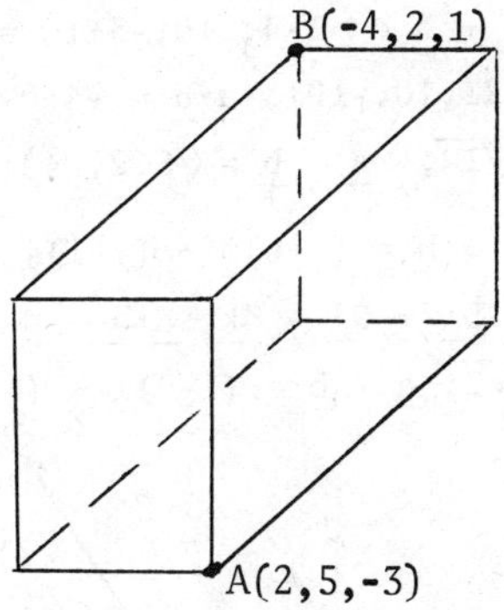

10. Since $d^2(A,B) = 6$, $d^2(B,C) = 33$, $d^2(A,C) = 27$, it is a right triangle with legs AB and AC, hypoteneuse BC. The area is $\frac{1}{2}d(A,B)d(A,C) = \frac{\sqrt{6}\sqrt{27}}{2} = \frac{(\sqrt{3}\sqrt{2})(3\sqrt{3})}{2} = \frac{9}{\sqrt{2}}$.

13. The given answer is immediate from (14.13) using $h = -5$, $k = 0$, $\ell = 1$, $r = \frac{1}{2}$.

16. The center will be the midpoint of AB, $(-3,5/2,0)$, and the radius is $\frac{1}{2}d(A,B) = \sqrt{89}/2$. Thus the equation: $(x+3)^2 + (y - 5/2)^2 + z^2 = 89/4$.

19. Completing the square: $(x^2-8x \quad) + y^2 + (z^2+8z \quad) = -16 \Longrightarrow$ $(x^2 - 8x + 16) + y^2 + (z^2 + 8z + 16) = 16 \Longrightarrow (x-4)^2 + y^2 + (z+4)^2 = 4^2$. Thus $C(4,0,-4)$, $r = 4$.

22. $x^2 + y^2 + z^2 - z = 0 \Longrightarrow x^2 + y^2 + (z - \frac{1}{2})^2 = \frac{1}{4}$. Thus $C(0,0,1/2)$, $r = \frac{1}{2}$.

26. $xyz = 0 \Longleftrightarrow$ at least one of x,y,z is 0. Thus the graph consists of the union of the 3 coordinate planes.

28. R is the set of points outside the sphere of radius 1 centered at the origin.

32. R is the set of points between the spheres of radii 2 and 3 centered at the origin.

34. R is the set of points above and below the line $y = x$ in the xy plane.

EXERCISES 14.3, page 675

1. (a) $\underline{a}\cdot\underline{a} = \langle a_1,a_2,a_3\rangle\cdot\langle a_1,a_2,a_3\rangle = a_1^{\,2} + a_2^{\,2} + a_3^{\,2} = |\underline{a}|^2$. The other parts follow from the definition and the following real number properties: $a_ib_i = b_ia_i$ for (b), $(ca_i)b_i = c(a_ib_i) = a_i(cb_i)$ for (d), and $0a_i = 0$ for (e), all of these holding true for $i = 1,2,3$.

4. $\underline{a} + \underline{b} = \langle 1-4,2+0,-3+1\rangle = \langle -3,2,-2\rangle$; $5\underline{a} - 4\underline{b} = \langle 5,10,-15\rangle - \langle -16,0,4\rangle = \langle 21,10,-19\rangle$; $\underline{a}\cdot\underline{b} = -4+0-3 = -7$; $|\underline{a}| = \sqrt{1+4+9} = \sqrt{14}$; $|3\underline{a}| = |-3\underline{a}| = 3|\underline{a}| = 3\sqrt{14}$; $\underline{a} - \underline{b} = \langle 5,2,-4\rangle$; $|\underline{a} - \underline{b}| = \sqrt{25 + 4 + 16} = \sqrt{45}$.

7. $\underline{a} + \underline{b} = (1+0)\underline{i} + (1-1)\underline{j} + (0+1)\underline{k} = \underline{i} + \underline{k}$; $5\underline{a} - 4\underline{b} = (5\underline{i} + 5\underline{j}) - (-4\underline{j} + 4\underline{k}) = 5\underline{i} + 9\underline{j} - 4\underline{k}$; $\underline{a}\cdot\underline{b} = 0-1+0 = 1$; $|\underline{a}| = \sqrt{1+1+0} = \sqrt{2}$; $|3\underline{a}| = |-3\underline{a}| = 3|\underline{a}| = 3\sqrt{2}$; $\underline{a} - \underline{b} = (1-0)\underline{i} + (1-(-1))\underline{j} + (0-1)\underline{k} = \underline{i} + 2\underline{j} - \underline{k}$, $|\underline{a}-\underline{b}| = \sqrt{1+4+1} = \sqrt{6}$.

10.

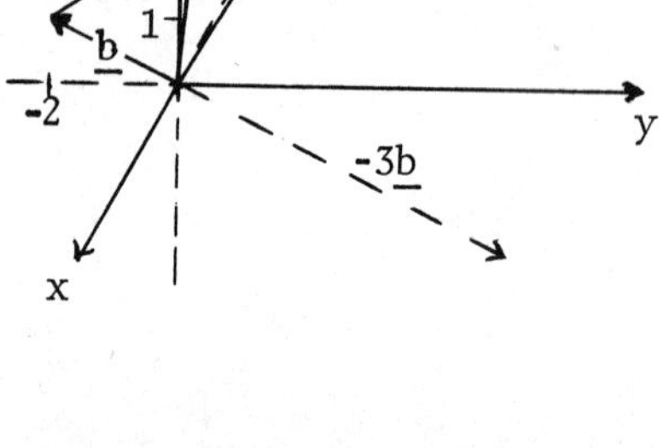

13. $\underline{a}\cdot(\underline{b}+\underline{c}) = \langle -2,3,1\rangle\cdot\langle 8,-1,7\rangle = -16-3+7 = -12$.

16. $(\underline{a}-\underline{b})\cdot(\underline{b}+\underline{c}) = \langle -9,-1,-4\rangle\cdot\langle 8,-1,7\rangle = -72+1-28 = -99$.

19. $\text{comp}_{\underline{b}}(\underline{a}+\underline{c}) = \dfrac{(\underline{a}+\underline{c})\cdot\underline{b}}{|\underline{b}|} = \dfrac{\langle -1,-2,3\rangle\cdot\langle 7,4,5\rangle}{\sqrt{49+16+25}} = \dfrac{-7-8+15}{\sqrt{90}} = 0$.

22. $\cos\theta = \dfrac{\underline{a} \cdot \underline{b}}{|\underline{a}|\ |\underline{b}|} = \dfrac{5 - 7(0) - 4}{\sqrt{66}\sqrt{26}} = \dfrac{1}{\sqrt{66}\sqrt{26}} \approx .024.$

25. $\overrightarrow{PQ} = \langle 1-3, 5-(-2), 4-(-1)\rangle = \langle -2,7,5\rangle$. $\overrightarrow{RS} = \langle -4-2, 1-0, 5-(-6)\rangle = \langle -6,1,11\rangle$.
$\overrightarrow{PQ} \cdot \overrightarrow{RS} = (-2)(-6) + 7(1) + 5(11) = 12 + 7 + 55 = 74.$

28. $\overrightarrow{QS} = \langle -4-1, 1-5, 5-4\rangle = \langle -5,-4,1\rangle$, and $|\overrightarrow{QS}| = \sqrt{25+16+1} = \sqrt{42}$. $\overrightarrow{RP} = \langle 3-2, -2-0, -1-(-6)\rangle = \langle 1,-2,5\rangle$, and $|\overrightarrow{RP}| = \sqrt{1+4+25} = \sqrt{30}$. If θ is the angle between $\overrightarrow{QS}$ and $\overrightarrow{RP}$, then $\cos\theta = \dfrac{\overrightarrow{QS} \cdot \overrightarrow{RP}}{|\overrightarrow{QS}|\ |\overrightarrow{RP}|} = \dfrac{-5+(-4)(-2)+5}{\sqrt{42}\sqrt{30}} = \dfrac{8}{6\sqrt{35}} \approx 0.2254.$

Thus $\theta = \cos^{-1}\left(\dfrac{8}{6\sqrt{35}}\right) \approx 1.3435$ or 76.98°.

31. $\overrightarrow{PR} = \langle x-8, y+3, z-5\rangle$, $\overrightarrow{PQ} = \langle -2,4,-12\rangle$. $\overrightarrow{PR}$ is orthogonal to $\overrightarrow{PQ}$ if $\overrightarrow{PR}\cdot\overrightarrow{PQ} = 0$, or $-2(x-8) + 4(y+3) - 12(z-5) = 0$.

34. From (14.26) and Example 6, we see that if a force is given by vector $\underline{a}$ and the path of motion by a vector $\underline{b}$, then the work done is $\underline{a}\cdot\underline{b}$. Here $\underline{b}$ is a vector corresponding to $\overrightarrow{PQ} = \langle 5,-1,5\rangle$ so that $\underline{a}\cdot\underline{b} = \langle 8,0,-4\rangle \cdot \langle 5,-1,-5\rangle = 40+0+20 = 60$.

35. Hint: If $\underline{F}$ is the force vector, then $\underline{F} = c\underline{a}$ for some $c > 0$, and $|\underline{F}| = 4$. Compute $|\underline{F}| = c|\underline{a}|$ to find c, then dot $\underline{F}$ with the vector from (0,2,0) to (0,-1,0).

37. Two nonzero vectors are orthogonal if their inner product is 0.
$(3\underline{i} - \underline{j} + c\underline{k})\cdot(2c\underline{i} + 3\underline{j} + 4\underline{k}) = 6c - 3 + 4c = 10c - 3 = 0 \iff c = 3/10.$

40. Let $\langle r,s\rangle$ be a unit vector orthogonal to $\langle -2,-1\rangle$. Then $r^2 + s^2 = 1$ and $\langle r,s\rangle \cdot \langle -2,-1\rangle = -2r-s = 0$. Thus $s = -2r$, $r^2 + (-2r)^2 = 5r^2 = 1 \Longrightarrow r = \pm 1/\sqrt{5}$ and $s = \mp 2/\sqrt{5}$. The 2 vectors are $\pm 1/\sqrt{5}\langle 1, -2\rangle$.

EXERCISES 14.4, page 679

NOTE: Since $|c\underline{a}| = |c|\,|\underline{a}|$, it is easy to show that the direction cosines of $\underline{a}$ and $c\underline{a}$ are the same if $c > 0$, whereas those of $c\underline{a}$ are the negatives of those of $\underline{a}$ if $c < 0$. Consider:

If $c > 0$, $\dfrac{1}{|c\underline{a}|}c\underline{a} = \dfrac{1}{c|\underline{a}|}c\underline{a} = \dfrac{1}{|\underline{a}|}\underline{a} = \langle \cos\alpha, \cos\beta, \cos\gamma\rangle$.

If $c < 0$, $\dfrac{1}{|c\underline{a}|}c\underline{a} = \dfrac{1}{(-c)|\underline{a}|}c\underline{a} = -\dfrac{1}{|\underline{a}|}\underline{a} = \langle -\cos\alpha, -\cos\beta, -\cos\gamma\rangle$.

1. By the above note, the direction cosines of $\underline{a}$ and $3\underline{a}$ are the same. Since $|\underline{a}| = \sqrt{30}$, those of $\underline{a}$ (and $3\underline{a}$) are $-2/\sqrt{30}$, $1/\sqrt{30}$, $5/\sqrt{30}$ by (14.28i).

4. Since $|\underline{a}| = \sqrt{26}$, the direction cosines of $\underline{a}$ and $3\underline{a}$ are $5/\sqrt{26}$, 0, $1/\sqrt{26}$, and $-2\underline{a}$ are $-\dfrac{5}{\sqrt{26}}$, 0, $-\dfrac{1}{\sqrt{26}}$.

7. $\overrightarrow{PQ} = \langle 3-7,2+2,-1-4\rangle = \langle -4,4,-5\rangle$, $|\overrightarrow{PQ}| = \sqrt{57}$ and the direction cosines are $-\frac{4}{\sqrt{57}}, \frac{4}{\sqrt{57}}, \frac{-5}{\sqrt{57}}$.

10. $\overrightarrow{PQ} = \langle 0,-4,0\rangle$, $|\overrightarrow{PQ}| = 4$ and the direction cosines are 0,-1,0.

13. $\cos^2\alpha + \cos^2\beta + \cos^2\gamma = 1 \Longrightarrow (1/3)^2 + (1/3)^2 + \cos^2\gamma = 1 \Longrightarrow \cos^2\gamma = 1 - 2/9 = 7/9 \Longrightarrow \cos\gamma = \sqrt{7}/3$.

16. Hint: Let $d_1 = (\ell_1{}^2 + m_1{}^2 + n_1{}^2)^{1/2}$, and d_2 similarly. Use #15a and results of this section to show that $\underline{a} = \frac{|\underline{a}|}{d_1}\langle \ell_1,m_1,n_1\rangle$ and similarly for $\underline{b}$. Now use (14.20) for (a) and the fact that the angle between parallel vectors is 0 or π for (b).

EXERCISES 14.5, page 682

1. A direction vector for the line is $\overrightarrow{P_1P_2} = \langle -3,8,-3\rangle$, and, using P_1 as the point, we obtain $x = 5-3t$, $y = -2+8t$, $z = 4-3t$. The xy-plane is intersected when $z = 0$, thus, at $t = 4/3$. Substituting we find $x = 5 - 3(\frac{4}{3}) = 1$, $y = -2 + 8(\frac{4}{3}) = \frac{26}{3}$, and the point of intersection is $(1,\frac{26}{3},0)$. Similarly the xz-plane is intersected when $y = 0$ at $t = 1/4$; the yz-plane is intersected when $x = 0$ and $t = 5/3$. These values of t yield the points in the answer.

4. $\overrightarrow{P_1P_2} = \langle 0,0,-7\rangle$ and, using P_1, the equations are $x = 2$, $y = -2$, $z = 4-7t$. The line is parallel to the z-axis and never intersects the xz- and yz-planes. The line pierces the xy-plane at (2,-2,0) when $t = 4/7$.

7. Here, as in #5, 6, and 8, the equations are immediate from (14.29).

10. Since $\overrightarrow{PQ} = \langle -3,8,-3\rangle$ and $\overrightarrow{RS} = \langle 10,10,-7\rangle$ are direction vectors for ℓ_1 and ℓ_2, the angle between ℓ_1 and ℓ_2 will be the angle between these vectors.

$$\cos\theta = \frac{\overrightarrow{PQ}\cdot\overrightarrow{RS}}{|\overrightarrow{PQ}|\,|\overrightarrow{RS}|} = \frac{71}{\sqrt{82}\,\sqrt{249}} \approx 0.497 \text{ and } \theta \approx 60.2^\circ.$$

13. If there is a point of intersection, that is, a point $P_o(x_o,y_o,z_o)$ simultaneously on both ℓ_1 and ℓ_2, then there is a value of t, say t_o, so that $1 + 2t_o = x_o$, $1 - 4t_o = y_o$, $5 - t_o = z_o$ (since $P_o \in \ell_1$) and a value of v, say v_o, so that $4 - v_o = x_o$, $-1 + 6v_o = y_o$, $4 + v_o = z_o$ (since $P_o \in \ell_2$). Equating the expressions for x_o,y_o,z_o we obtain the following system of 3 equations in 2 unknowns.

$$\left.\begin{aligned} 1 + 2t_o &= 4 - v_o \\ 1 - 4t_o &= -1 + 6v_o \\ 5 - t_o &= 4 + v_o \end{aligned}\right\} \Longrightarrow \left\{\begin{aligned} 2t_o + v_o &= 3 \\ -4t_o - 6v_o &= -2 \\ -t_o - v_o &= -1 \end{aligned}\right.$$

Solving the 1st and 3rd, by adding them, we get $t_o = 2$, $v_o = -1$ which also satisfy the 2nd equation. This is then the unique solution. Using either $t_o = 2$ in the ℓ_1 equations or $v_o = -1$ in the ℓ_2 equations the point of intersection (5,-7,3) results.

16. Proceeding as in #13, we get the system

$$\left.\begin{aligned} 2 - 5t_o &= 4 - 3v_o \\ 6 + 2t_o &= 7 + 5v_o \\ -3 - 2t_o &= 1 + 4v_o \end{aligned}\right\} \Longrightarrow \left\{\begin{aligned} 5t_o - 3v_o &= -2 \\ 2t_o - 5v_o &= 1 \\ 2t_o + 4v_o &= -4 \end{aligned}\right.$$

The 2nd and 3rd equations have the unique solution $t_o = -8/9$, $v_o = -5/9$. But these do not satisfy the 1st equation since $5t_o - 3v_o$ works out to -25/9, not -2. Thus the system has no solution, and the lines do not intersect.

19. From the figure, $d = \sqrt{|\overrightarrow{AB}|^2 - |\overrightarrow{BD}|^2}$. Now, $|\overrightarrow{AB}| = \sqrt{110}$ and $|\overrightarrow{BD}| = \text{comp}_{\overrightarrow{BC}}\overrightarrow{BA} = \dfrac{\overrightarrow{BC}\cdot\overrightarrow{BA}}{|\overrightarrow{BC}|} = \dfrac{67}{\sqrt{90}}$. Thus

$$d = \sqrt{110 - \frac{67^2}{90}} = \sqrt{\frac{5411}{90}}.$$

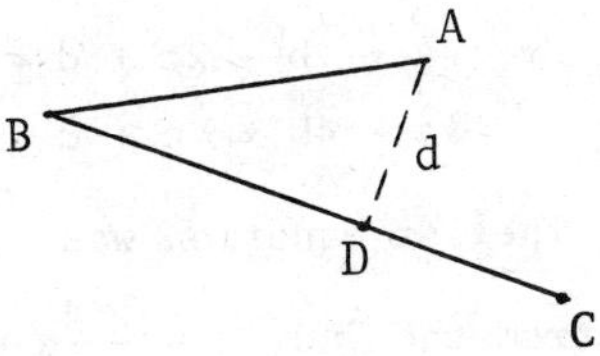

EXERCISES 14.6, page 688

1. (a) (b) (c)

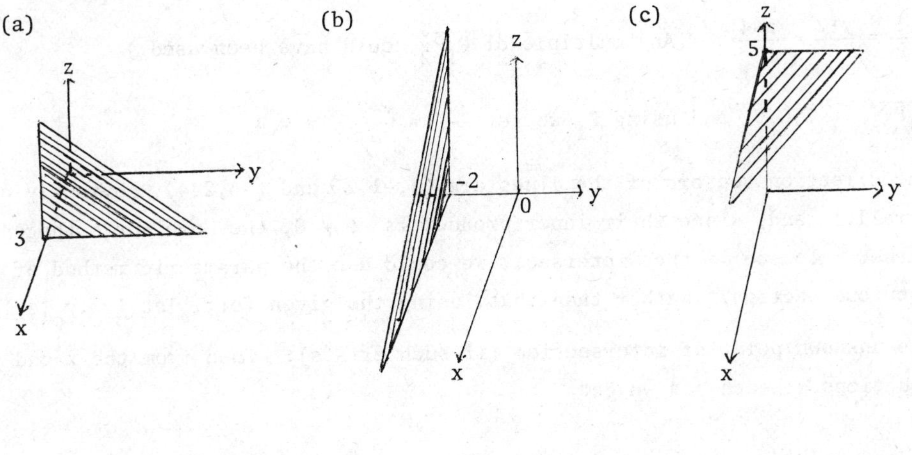

4.

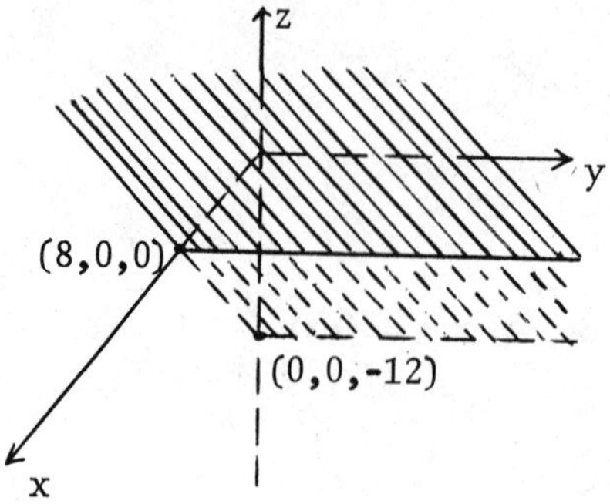

7.

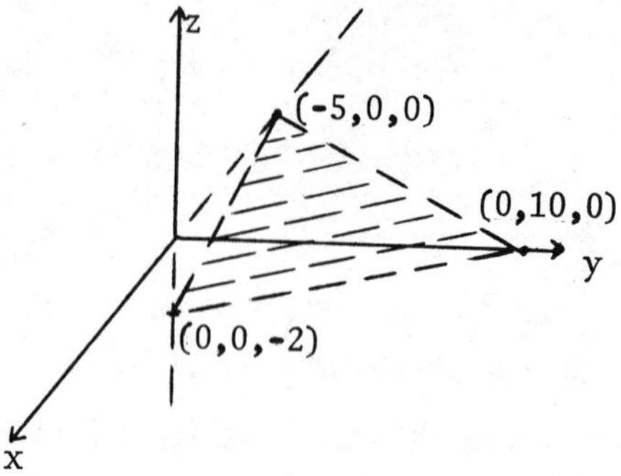

10. Using (14.30), we quickly get: (a) $x+2 = 0$, (b) $y-5 = 0$, (c) $z+8 = 0$.

13. Parallel planes have parallel normal vectors. Thus the desired plane has normal $\underline{n}$ which is a multiple of $\langle 3,-1,2\rangle$. The easiest way is to choose $\underline{n} = \langle 3,-1,2\rangle$ and the answer is obtained from (14.30).

16. The desired equation is of the form $ax + by + cz + d = 0$. Substituting the coordinates of P, Q, and R we obtain the system of 3 equations in 4 unknowns:

$$\left.\begin{aligned} 3a + 2b + c + d &= 0 \\ -a + b - 2c + d &= 0 \\ 3a - 4b + c + d &= 0 \end{aligned}\right\} \implies \left\{\begin{aligned} 3a + 2b + c + d &= 0 \\ -4a - b - 3c &= 0 \\ -6b &= 0 \end{aligned}\right.$$

(The 1st equation was subtracted from the others.) From the 3rd, $b = 0$ and from the 2nd, $c = -\frac{4}{3}a$ which produces from the 1st: $d = -3a - c = -3a + \frac{4}{3}a = -\frac{5}{3}a$. Selecting $a = -3$ we get $c = 4$, $d = 5$, and the equation is $-3x + 4z + 5 = 0$. (Any other choice for a would result in a multiple of this equation.)

19. $\overrightarrow{P_1P_2} = \langle -3,8,-3\rangle$ is a direction vector for the line, and using P_1 we obtain: $\frac{x-5}{-3} = \frac{y+2}{8} = \frac{z-4}{-3}$. (Any multiple of $\overrightarrow{P_1P_2}$ could have been used.)

22. $\overrightarrow{P_1P_2} = \langle -7,6,0\rangle$ and using P_1 we get $\frac{x-5}{-7} = \frac{y+7}{6}$, $z = 4$.

25. The direction vectors of the lines are $\langle 3,-1,2\rangle$ and $\langle -4,2,4\rangle$, obviously not parallel, and, since their inner product is $-6 \neq 0$, they are not orthogonal either. To see if they intersect, we could use the parametric method of the previous section. Rather than that, using the given form, let (x_o,y_o,z_o) be the unknown point of intersection (if such exists). Then from the x and y equations of each set we get

$$\left.\begin{aligned}\frac{x_o+1}{3} &= \frac{y_o-3}{-1}\\ \frac{x_o}{-4} &= \frac{y_o-4}{2}\end{aligned}\right\} \Longrightarrow \left.\begin{aligned} x_o &= -3y_o + 8\\ x_o &= -2y_o + 8\end{aligned}\right\} \Longrightarrow y_o = 0,\ x_o = 8$$

Using these values in each set, we find $z_o = 1$ in both cases. Thus the lines intersect at (8,0,1).

28. Using the formula of Example 7,

$$h = \frac{|2(3) + 4(1) - 5(-2) + 1|}{\sqrt{4 + 16 + 25}} = \frac{21}{\sqrt{45}}.$$

31. The plane containing ℓ contains the point Q(1,4,-3), obtained by setting t = 0. The desired normal vector, **n**, must be orthogonal to $\overrightarrow{PQ} = \langle -4,4,-5\rangle$ (a directed line segment in the plane) and also orthogonal to $\mathbf{a} = \langle 3,-2,1\rangle$ (the direction vector of ℓ). If $\mathbf{n} = \langle a,b,c\rangle$ we obtain from (14.20),

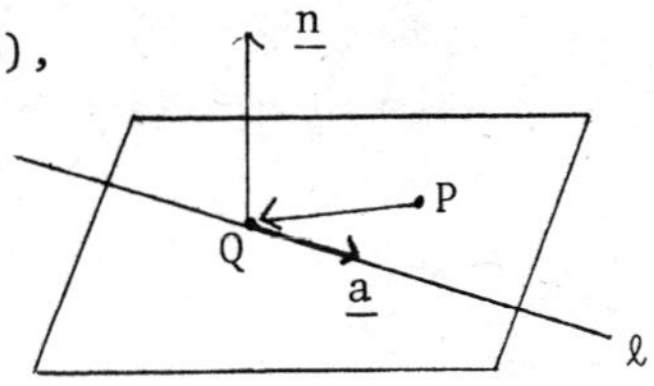

$$\left.\begin{aligned}\mathbf{n}\cdot\mathbf{a} &= 3a - 2b + c = 0\\ \mathbf{n}\cdot\overrightarrow{PQ} &= -4a + 4b - 5c = 0\end{aligned}\right\} \Longrightarrow \left\{\begin{aligned} 3a - 2b + c &= 0\\ 2a \qquad - 3c &= 0\end{aligned}\right.$$

where we added 2 times the first equation to the second. Any solution of this system will serve to get **n**. Selecting a = 6 and c = 4 to satisfy the 2nd equation, we get from the 1st, b = (3a+c)/2 = (18+4)/2 = 11. Thus one choice for **n** is ⟨6,11,4⟩ (there are infinitely many others; all are multiples of this choice). Using P(5,0,2), the equation is 6(x-5) + 11(y-0) + 4(z-2) = 0 or 6x + 11y + 4z - 38 = 0.

EXERCISES 14.7, page 698

1.
$$\mathbf{a}\times\mathbf{b} = \begin{vmatrix}\mathbf{i} & \mathbf{j} & \mathbf{k}\\ 1 & -2 & 3\\ 2 & 1 & -4\end{vmatrix} = \begin{vmatrix}-2 & 3\\ 1 & -4\end{vmatrix}\mathbf{i} - \begin{vmatrix}1 & 3\\ 2 & -4\end{vmatrix}\mathbf{j} + \begin{vmatrix}1 & -2\\ 2 & 1\end{vmatrix}\mathbf{k} =$$

$(8-3)\mathbf{i} - (-4-6)\mathbf{j} + (1+4)\mathbf{k} = 5\mathbf{i} + 10\mathbf{j} + 5\mathbf{k} = \langle 5,10,5\rangle$.

4.
$$\mathbf{a}\times\mathbf{b} = \begin{vmatrix}\mathbf{i} & \mathbf{j} & \mathbf{k}\\ 2 & 1 & 0\\ 0 & -5 & 2\end{vmatrix} = (2-0)\mathbf{i} - (4-0)\mathbf{j} + (-10-0)\mathbf{k} = 2\mathbf{i} - 4\mathbf{j} - 10\mathbf{k}.$$

7.
$$\underline{a}\times\underline{b}=\begin{vmatrix}\underline{i} & \underline{j} & \underline{k}\\ 3 & -1 & 8\\ 0 & 5 & 0\end{vmatrix} = (0-40)\underline{i} - (0-0)\underline{j} + (15-0)\underline{k} = -40\underline{i} + 15\underline{k}.$$

10.
$$\underline{a}\times\underline{b}=\begin{vmatrix}\underline{i} & \underline{j} & \underline{k}\\ 3 & 0 & 0\\ 0 & 0 & 4\end{vmatrix} = (0-0)\underline{i} - (12-0)\underline{j} + (0-0)\underline{k} = -12\underline{j}.$$

13. Hint: Evaluate the determinants in (14.37) and dot the result with $\underline{b}$.

16. Hint: Use (14.43, (iii) and (iv)) and the fact that $\underline{a}\times\underline{a}=\underline{0}$ for all vectors $\underline{a}$.

19. (a) First, $\overrightarrow{PR}\times\overrightarrow{PQ}$ is perpendicular to both $\overrightarrow{PR}$ and $\overrightarrow{PQ}$ and, therefore, to the plane of P, Q, and R. Since $\overrightarrow{PR}=\langle 2,-3,-1\rangle$, $\overrightarrow{PQ}=\langle -1,4,-3\rangle$, we get

$$\overrightarrow{PR}\times\overrightarrow{PQ}=\begin{vmatrix}\underline{i} & \underline{j} & \underline{k}\\ 2 & -3 & -1\\ -1 & 4 & -3\end{vmatrix} = (9+4)\underline{i} - (-6-1)\underline{j} + (8-3)\underline{k} = 13\underline{i} + 7\underline{j} + 5\underline{k}.$$ Any

multiple of this would also serve.

(b) Using P(1,-1,2) and the normal vector from part (a), we obtain: $13(x-1) + 7(y+1) + 5(z-2) = 0$.

(c) Since $|\overrightarrow{PR}\times\overrightarrow{PQ}| = \sqrt{169+49+25} = \sqrt{243} = 9\sqrt{3}$, the area of the triangle determined by P, Q, R is $9\sqrt{3}/2$.

22. (a)
$$\overrightarrow{PR}\times\overrightarrow{PQ}=\begin{vmatrix}\underline{i} & \underline{j} & \underline{k}\\ 6 & -2 & 1\\ 1 & 0 & -3\end{vmatrix} = 6\underline{i} + 19\underline{j} + 2\underline{k}.$$

(b) Using P(-1,2,0) and part (a), we get $6(x+1) + 19(y-2) + 2(z-0) = 0$, or $6x + 19y + 2z = 32$.

(c) Area $= \frac{1}{2}|\overrightarrow{PR}\times\overrightarrow{PQ}| = \frac{1}{2}\sqrt{36+361+4} = \frac{1}{2}\sqrt{401}$.

25. With $\underline{a}=\overrightarrow{OP}=\langle 1,-1,2\rangle$, $\underline{b}=\overrightarrow{OQ}=\langle 0,3,-1\rangle$, $\underline{c}=\overrightarrow{OR}=\langle 3,-4,1\rangle$, the volume is $|\underline{a}\cdot(\underline{b}\times\underline{c})| =$

$$\left|\begin{vmatrix}1 & -1 & 2\\ 0 & 3 & -1\\ 3 & -4 & 1\end{vmatrix}\right| = |(1)(3-4)-(-1)(0+3)+2(0-9)| = |-1+3-18| = 16.$$

28. Using Example 7 with $A = P_1$, $B = Q_1$, $C = P_2$, $D = Q_2$, the formula is

$$d = \frac{1}{|\overrightarrow{AB}\times\overrightarrow{CD}|}\,|(\overrightarrow{AB}\times\overrightarrow{CD})\cdot\overrightarrow{AC}|.$$

Now, $\overrightarrow{AB}=\langle -1,1,5\rangle$, $\overrightarrow{CD}=\langle 7,2,-2\rangle$ yield

$$\overrightarrow{AB} \times \overrightarrow{CD} = \begin{vmatrix} \underline{i} & \underline{j} & \underline{k} \\ -1 & 1 & 5 \\ 7 & 2 & -2 \end{vmatrix} = (-2-10)\underline{i} - (2-35)\underline{j} + (-2-7)\underline{k} = -3(4\underline{i} - 11\underline{j} + 3\underline{k})$$

$\Longrightarrow$ $|\overrightarrow{AB} \times \overrightarrow{CD}| = 3\sqrt{16+121+9} = 3\sqrt{146}$. Next, $\overrightarrow{AC} = \langle -3,-4,2 \rangle$ so that $(\overrightarrow{AB} \times \overrightarrow{CD}) \cdot \overrightarrow{AC} = (-12)(-3) + (33)(-4) + (-9)(2) = 36 - 132 - 18 = -114$. Thus $d = \dfrac{|-114|}{3\sqrt{146}} = \dfrac{38}{\sqrt{146}} \approx 3.145$.

EXERCISES 14.8, page 704

4.

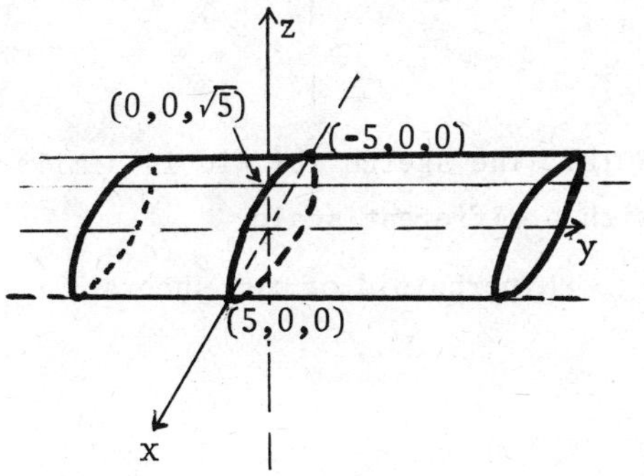

10.

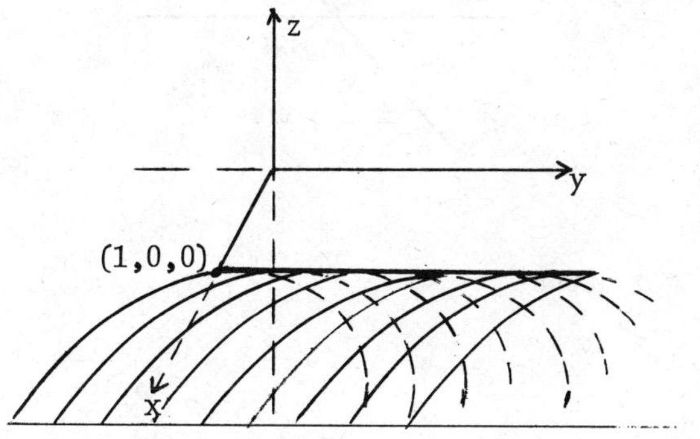

13. Replacing y by $\pm\sqrt{x^2+y^2}$, we get $z = 4 - (\pm\sqrt{x^2+y^2})^2$ or $z = 4-x^2-y^2$.

16. Replacing x by $\pm\sqrt{y^2+x^2}$, we get $\pm\sqrt{y^2+x^2}\ z = 1$ or $(y^2+x^2)z^2 = 1$.

13.

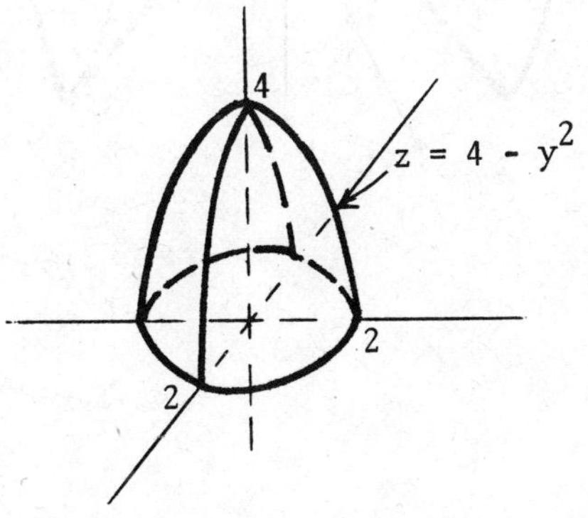

16.

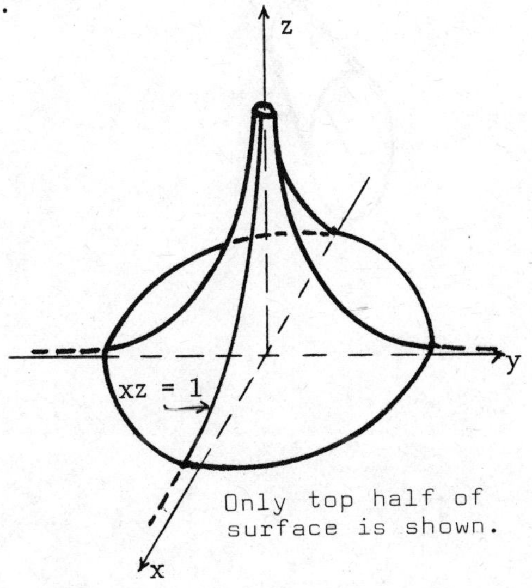

EXERCISES 14.9, page 710

1. Paraboloid

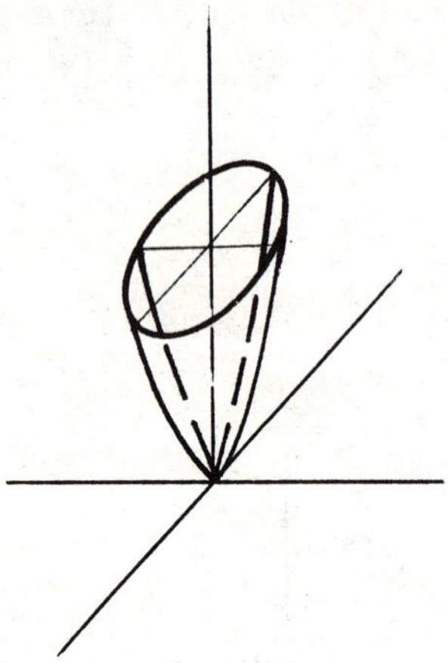

4. Hyperboloid of one sheet

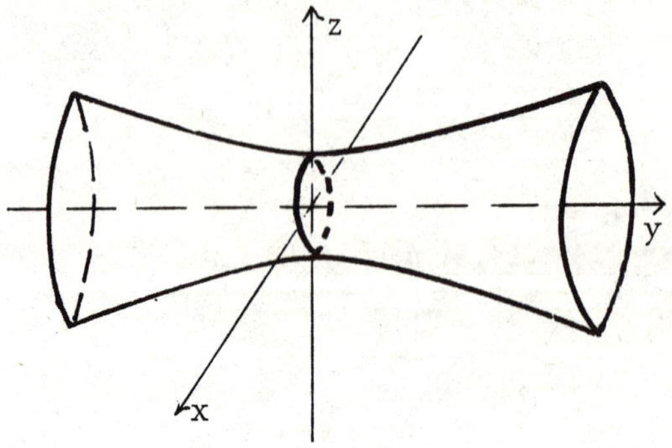

NOTE: The sketch for #10 is similar but with a different scale.

7. Cone

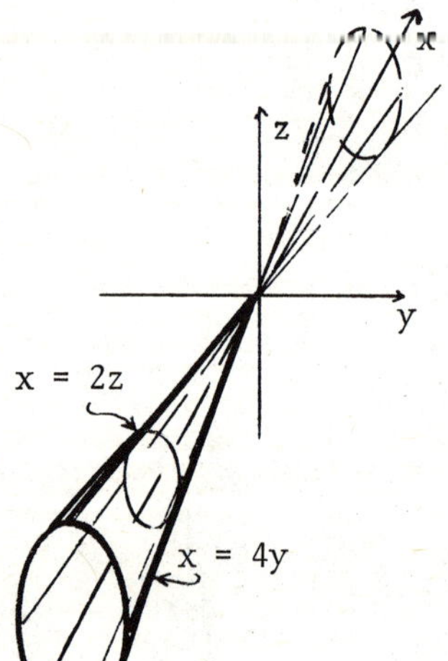

13. Hyperboloid of two sheets

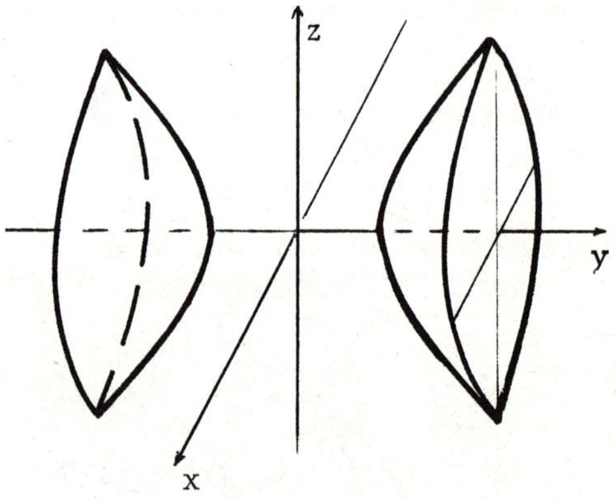

16. Paraboloid

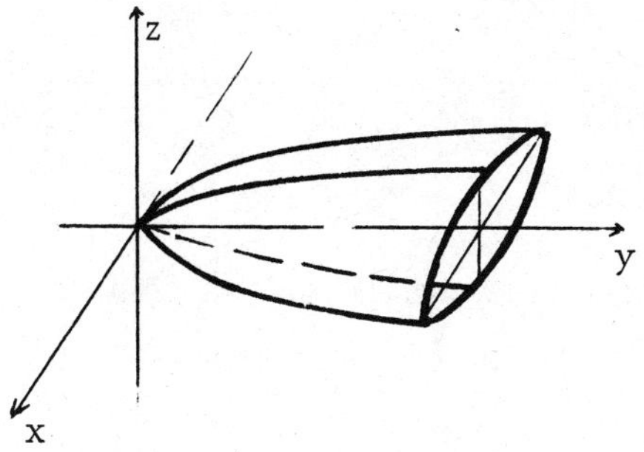

19. Hyperboloic paraboloid

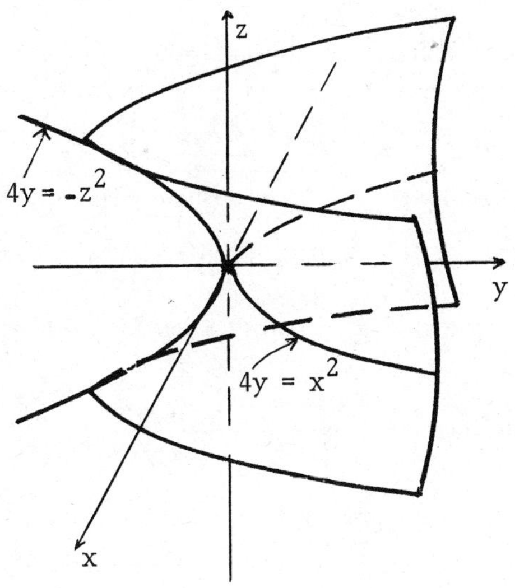

EXERCISES 14.10, page 713

1. (a) $x = 5\cos\frac{\pi}{2} = 0$, $y = 5\sin\frac{\pi}{2} = 5$, $z = 3$.

(b) $x = 6\cos\frac{\pi}{3} = 3$, $y = 6\sin\frac{\pi}{3} = 3\sqrt{3}$, $z = -5$.

4. (a) $r = \sqrt{(2\sqrt{3})^2+2^2} = 4$, $\tan\theta = \dfrac{2}{2\sqrt{3}} = \dfrac{1}{\sqrt{3}}$ and θ can be chosen to be $\frac{\pi}{6}$ (since x and y are positive), $z = -2$.

(b) $r = \sqrt{2+2} = 2$, $\tan\theta = \dfrac{-\sqrt{2}}{\sqrt{2}} = -1$, and θ can be chosen to be $-\frac{\pi}{4}$ (since $x > 0$, $y < 0$), $z = 1$.

7. $r = 4 \Longrightarrow r^2 = 16 \Longrightarrow x^2 + y^2 = 16$, a right circular cylinder of radius 4 whose axis is the z-axis.

10. $\rho = 3 \Longrightarrow \rho^2 = 9 \Longrightarrow x^2+y^2+z^2 = 9$, a sphere of radius 3 centered at the origin.

13. $\rho = 4\cos\phi \Longrightarrow \rho^2 = 4\rho\cos\phi \Longrightarrow x^2+y^2+z^2 = 4z \Longrightarrow x^2+y^2+z^2-4z+4 = 4 \Longrightarrow x^2+y^2+(z-2)^2 = 4$, a sphere of radius 2 centered at the point (0,0,2).

16. The graph of $\phi = \pi/2$ is the xy-plane since ϕ, the angle between $\overrightarrow{OP}$ and the positive z-axis, is $\pi/2$ if, and only if, P is in the xy-plane.

19. $\rho^2-3\rho+2 = 0 \Longleftrightarrow (\rho-2)(\rho-1) = 0 \Longrightarrow \rho = 1$ or 2. Thus, the graph is the union of the two spheres of radii 1 and 2 centered at the origin.

22. Cylindrical: $x^2+y^2 = 4z \Longrightarrow r^2 = 4z$.

Spherical: $x^2+y^2 = 4z \Longrightarrow \rho^2\sin^2\phi\cos^2\theta + \rho^2\sin^2\phi\sin^2\theta = 4\rho\cos\phi \Longrightarrow$
$\rho^2\sin^2\phi = 4\rho\cos\phi \Longrightarrow \rho = 4\cot\phi\csc\phi$.

25. Cylindrical: $y^2+z^2 = 9 \Longrightarrow r^2\sin^2\theta + z^2 = 9$.
Spherical: $y^2+z^2 = 9 \Longrightarrow \rho^2\sin^2\phi\sin^2\theta + \rho^2\cos^2\phi = 9$.

EXERCISES 14.11 (Review), page 713

1. $4\underline{a} + \underline{b} = (8\underline{i} + 20\underline{j}) + (4\underline{i} - \underline{j}) = 12\underline{i} + 19\underline{j}$.

4. $3\underline{a} \cdot (5\underline{b} + \underline{i}) = (6\underline{i} + 15\underline{j}) \cdot (21\underline{i} - 5\underline{j}) = 126 - 75 = 51$.

7. Since $\underline{a}$ "points" into the 1st quadrant, its direction is $\tan^{-1}\frac{5}{2} \approx 68.2^{\circ}$.

10. $\cos\theta = \dfrac{\underline{a}\cdot\underline{b}}{|\underline{a}||\underline{b}|} = \dfrac{8-5}{\sqrt{29}\sqrt{17}} = \dfrac{3}{\sqrt{493}} \approx 0.135$.

13. The form of the equation is $ax + by + cz + d = 0$. Substituting the coordinates of the intercepts $(5,0,0)$, $(0,-2,0)$, $(0,0,6)$ we get $5a + d = 0$, $-2b + d = 0$, $6c + d = 0$. Choosing $d = -1$, we obtain $a = 1/5$, $b = -1/2$, $c = 1/6$ yielding $\frac{x}{5} - \frac{y}{2} + \frac{z}{6} - 1 = 0$ or $6x - 15y + 5z = 30$.

16. Replacing x by $+\sqrt{y^2+x^2}$ we get $z = +\sqrt{y^2+x^2}$ or $z^2 = x^2+y^2$.

19.

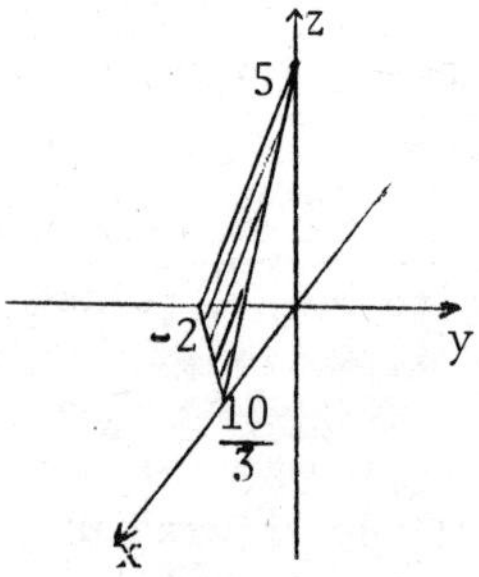

22.

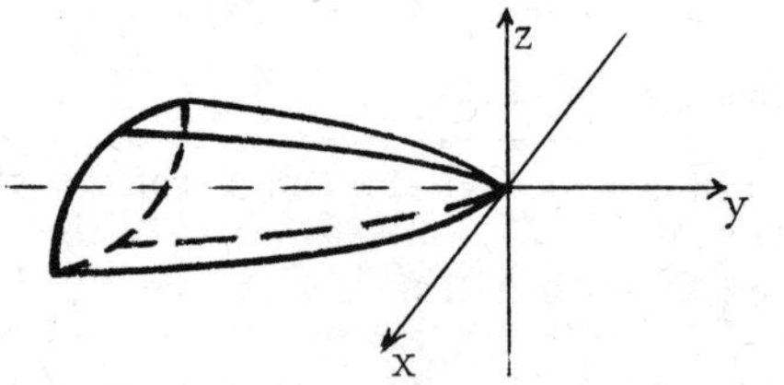

25.

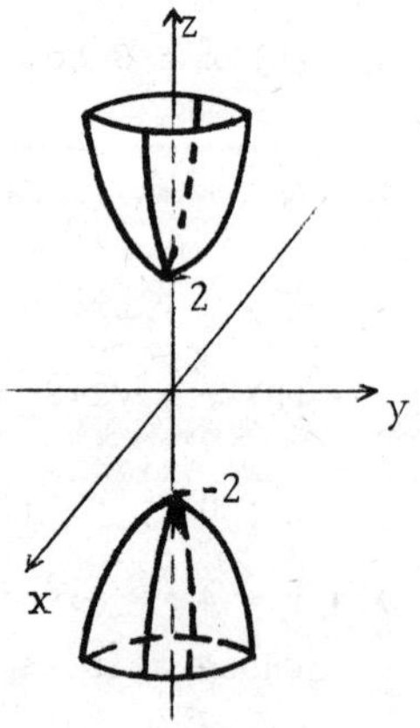

28. $3\underline{a} - 2\underline{b} = (9\underline{i} - 3\underline{j} - 12\underline{k}) - (4\underline{i} + 10\underline{j} - 4\underline{k}) = 5\underline{i} - 13\underline{j} - 8\underline{k}$.

31. $|\underline{b}| + |\underline{c}| = \sqrt{4+25+4} + \sqrt{1+0+36} = \sqrt{33} + \sqrt{37}$.

34. $\cos\theta = \dfrac{3(-1)-1(0)-4(6)}{\sqrt{26}\sqrt{37}} = \dfrac{-27}{\sqrt{26}\sqrt{37}}$.

37. $\text{comp}_{\underline{b}}\,\underline{a} = \dfrac{\underline{a}\cdot\underline{b}}{|\underline{b}|} = \dfrac{6-5+8}{\sqrt{33}} = \dfrac{9}{\sqrt{33}}$.

40. For any vector $\underline{a}$, $|\underline{a} \times \underline{a}| = |\underline{a}|^2\sin 0 = 0 \Longrightarrow \underline{a} \times \underline{a} = \underline{0}$. (Direct computation

gives the same result.

43. $-2\underline{b}$ is the desired vector by (14.6ii) and (14.7). $-2\underline{b} = -2(2\underline{i}+5\underline{j}-2\underline{k}) = -4\underline{i} - 10\underline{j} + 4\underline{k}$.

46. The 1st line has direction vector $\underline{a} = 2\underline{i} - 4\underline{j} + 8\underline{k}$. But to get one for the 2nd line, we must rewrite the equations as: $\frac{x+1}{7} = \frac{y-6}{-2} = \frac{(z + 7/2)}{-2}$. Now, from this form, we see that a direction vector is $\underline{b} = 7\underline{i} - 2\underline{j} - 2\underline{k}$ (NOT $7\underline{i} + 2\underline{j} - 4\underline{k}$ as you might have guessed from a first glance at the original equations). As usual, $\cos\theta = \frac{\underline{a}\cdot\underline{b}}{|\underline{a}||\underline{b}|} = \frac{14 + 8 - 16}{\sqrt{84}\,\sqrt{57}} = \frac{1}{\sqrt{133}} \approx 0.087$ and $\theta \approx 85^{\circ}$.

49. The direction vectors are $\langle 1,1,7\rangle$ and $\langle 5,-2,-4\rangle$, and $\cos\theta = \frac{5-2-28}{\sqrt{51}\,\sqrt{45}} \approx -0.522$, $\theta \approx 121.5^{\circ}$.

52. Cylindrical: $r = \rho\sin\phi = 12\sin\frac{\pi}{6} = 6$, $\theta = 3\pi/4$ (given),

$z = \rho\cos\phi = 12\cos\frac{\pi}{6} = 6\sqrt{3}$.

Rectangular: (using the cylindrical coordinates just obtained)

$x = r\cos\theta = 6\cos\frac{3\pi}{4} = -\frac{6}{\sqrt{2}}$, $y = r\sin\theta = 6\sin\frac{3\pi}{4} = \frac{6}{\sqrt{2}}$, $z = 6\sqrt{3}$.

55. In spherical coordinates $x = \rho\sin\phi\cos\theta$, and thus the equation is $x = 1$, a plane.

58. Cylindrical: $z = x^2-y^2 = r^2\cos^2\theta - r^2\sin^2\theta = r^2\cos 2\theta$.

Spherical: $z = x^2-y^2 \Longrightarrow \rho\cos\phi = \rho^2\sin^2\phi\cos^2\theta - \rho^2\sin^2\phi\sin^2\theta \Longrightarrow$

$\cos\phi = \rho\sin^2\phi\cos 2\theta$.

CHAPTER 15

VECTOR-VALUED FUNCTIONS

EXERCISES 15.1, page 720

NOTE: In this section we list, as we have previously, the integral answers for the odd-numbered problems whose solutions are not included.

1. C has parametric equations $x = t$, $y = 4 \cos t$, $z = 9 \sin t$. Since $(\frac{y}{4})^2 + (\frac{z}{9})^2 = \cos^2 t + \sin^2 t = 1$, the curve is a helix lying on the elliptic cylinder $\frac{y^2}{16} + \frac{z^2}{81} = 1$.

4. C has parametric equations $x = t^3$, $y = t^2$, $z = t$. Since $t = x^{1/3}$ and $y = t^2 = x^{2/3}$, the curve lies above the curve $y = x^{2/3}$ in the xy-plane.

4. (Graph)

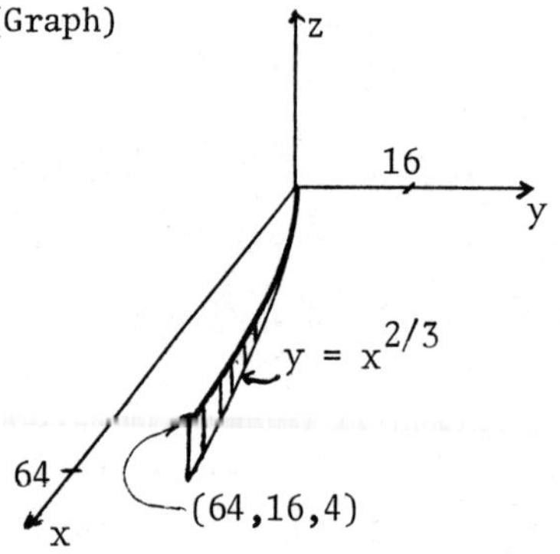

10. (Graph)

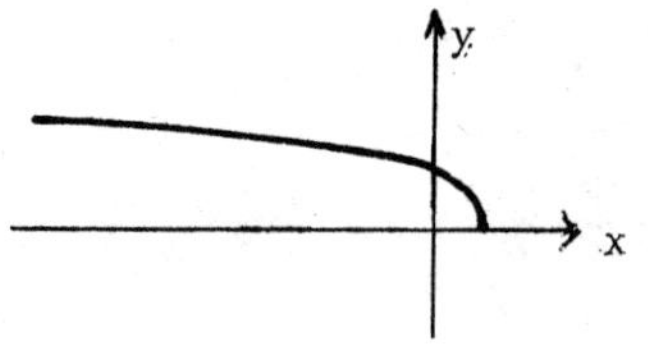

7. C has parametric equations $x = t$, $y = t$, $z = \sin t$. The curve, then, is the graph of $z = \sin t = \sin x$ in the plane $y = x$.

10. C is a curve in the xy-plane with parametric equations $x = 1-t^3$, $y = t$, $t \geq 0$. Eliminating t, we get $x = 1-y^3$, or $y = \sqrt[3]{1-x}$, $y \geq 0$. (The graph is next to that of #4 above.)

13. $x'^2 + y'^2 + z'^2 = 25 + 64t^2 + 36t^2 = 25 + 100t^2 = 25(1+4t^2)$. Thus, by (15.2),

$L = 5\int_0^2 \sqrt{1+4t^2}\, dt$. Let $u = 2t$ so that $dt = du/2$. $t = 0,2 \Rightarrow u = 0,4$.

Making these changes and using Formula 21 in the Table of Integrals,

$$L = \frac{5}{2}\int_0^4 \sqrt{1+u^2}\, du = \frac{5}{2}\Big[\frac{u}{2}\sqrt{1+u^2} + \frac{1}{2}\ln|u + \sqrt{1+u^2}|\Big]_0^4 =$$

$\frac{5}{2}[(\frac{4}{2}\sqrt{17} + \frac{1}{2}\ln|4 + \sqrt{17}|) - (0 + \frac{1}{2}\ln|0+1|)]$. Since $\ln 1 = 0$,

$$L = \frac{5}{4}(4\sqrt{17} + \ln(4 + \sqrt{17})) \approx 23.234.$$

15. $L = \int_0^{2\pi} \sqrt{(\cos t - \sin t)^2 e^{2t} + e^{2t} + (\cos t + \sin t)^2 e^{2t}}\, dt.$

16. $x'^2 + y'^2 + z'^2 = 4 + 144\cos^2 3t + 144\sin^2 3t = 4 + 144(\sin^2 3t + \cos^2 3t)$

$= 4 + 144(1) = 148.$ Thus $L = \int_0^{2\pi} \sqrt{148}\,dt = 2\pi\sqrt{148} = 4\pi\sqrt{37} \approx 76.44.$

17. $L = \int_0^1 \sqrt{36t^2 + 9t^4 + 36}\,dt.$

EXERCISES 15.2, page 727

1. (a) The domain of $\sqrt{t-1}$ is $[1,\infty)$, and that of $\sqrt{2-t}$ is $(-\infty,2]$. The domain of $\underline{r}$ is the intersection of these intervals, $[1,2]$. $\underline{r}$ is continuous there since both component functions are. (b) The answers given for $\underline{r}'$ and $\underline{r}''$ follow from $D_t\sqrt{t-1} = \frac{1}{2}(t-1)^{-1/2}$, $D_t^2\sqrt{t-1} = -\frac{1}{4}(t-1)^{-3/2}$ and $D_t\sqrt{2-t} = -\frac{1}{2}(2-t)^{-1/2}$, $D_t^2\sqrt{2-t} = -\frac{1}{4}(2-t)^{-3/2}$.

4. (a) e^{t^2} has domain $(-\infty,\infty)$; $\sin^{-1}t$ has domain $[-1,1]$. Thus the domain of $\underline{r}$ is $[-1,1]$ where $\underline{r}$ is continuous. (b) $\underline{r}'(t) = \langle 2te^{t^2}, 1/\sqrt{1-t^2}\rangle$, $\underline{r}''(t) = \langle (4t^2+2)e^{t^2}, t/\sqrt{(1-t^2)^3}\rangle$.

7. (a) The curve has parametric equations $x = 4\cos t$, $y = 2\sin t$. Thus $\frac{x}{4} = \cos t$, $\frac{y}{2} = \sin t$ and, since $\cos^2 t + \sin^2 t = 1$, the curve is the ellipse $\frac{x^2}{16} + \frac{y^2}{4} = 1.$

(b) $\underline{r}'(t) = -4\sin t\,\underline{i} + 2\cos t\,\underline{j}$,

$\underline{r}(\frac{3\pi}{4}) = -\frac{4}{\sqrt{2}}\underline{i} + \sqrt{2}\underline{j}$, $\underline{r}'(\frac{3\pi}{4}) = -\frac{4}{\sqrt{2}}\underline{i} - \frac{2}{\sqrt{2}}\underline{j}$.

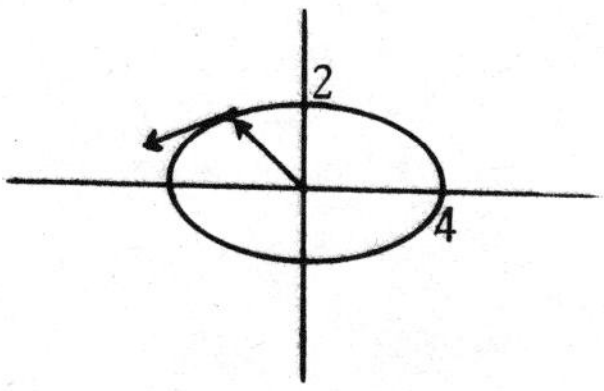

10. (a) $y = t^3 \Longrightarrow y^{2/3} = t^2 = x \Longrightarrow$ the desired curve is the graph of $x = y^{2/3}$.
(b) $\underline{r}'(t) = 2t\,\underline{i} + 3t^2\,\underline{j}$. $\underline{r}(-1) = \underline{i} - \underline{j}$ and $\underline{r}'(-1) = -2\underline{i} + 3\underline{j}$.

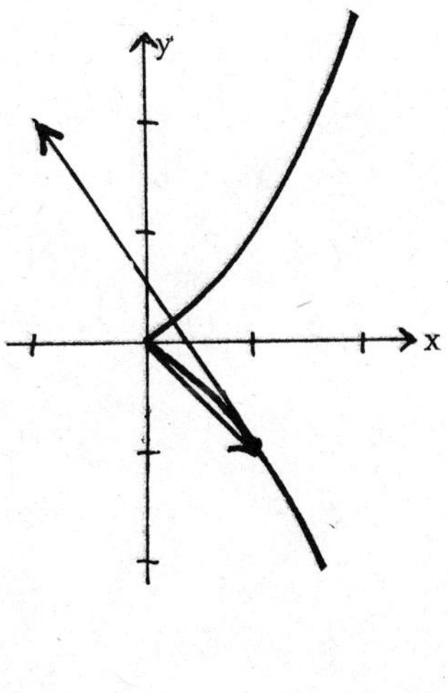

13. (a) Tan t has domain $\{t: t \neq \frac{\pi}{2} + n\pi\}$ where n is an integer. The other functions have domain $(-\infty,\infty)$. Thus the domain of $\underline{r}$, being the intersection of the domains of all of its component functions, is the same as that of tan t. (b) $\underline{r}'(t) = 2t\,\underline{i} + \sec^2 t\,\underline{j} + 0\,\underline{k}$, $\underline{r}''(t) = 2\underline{i} + 2\sec t\, D_t(\sec t)\underline{j} + 0\,\underline{k}$.

16. (a) The domain of $\ln(1-t)$ is $(-\infty,1)$ and the other functions have domain $(-\infty,\infty)$. Thus the domain of $\underline{r}$ is $(-\infty,1)$ and $\underline{r}$ is continuous there since all the components are.
(b) $\underline{r}'(t) = -(1-t)^{-1}\underline{i} + \cos t\,\underline{j} + 2t\,\underline{k}$, $\underline{r}''(t) = -(1-t^{-2})\underline{i} - \sin t\,\underline{j} + 2\underline{k}$.

19. The point P(1,0,4) occurs when t = 0. If $\underline{r}(t) = \langle e^t, te^t, t^2+4\rangle$, then $\underline{r}'(t) = \langle e^t, (t+1)e^t, 2t\rangle$, and $\underline{r}'(0) = \langle 1,1,0\rangle$ is a tangent vector to C at P and, hence, a direction vector for the tangent line. By (14.29) parametric equations for the tangent line are $x = 1+t$, $y = 0+t$, $z = 4+0t$.

22. We may think of the curve as determined by the vector function $\underline{r}(t) = \langle 2 + \sin t, \cos t, t\rangle$, and the point P(2,1,0) occurs when t = 0. Now, $\underline{r}'(t) = \langle \cos t, -\sin t, 1\rangle$, and $\underline{r}'(0) = \langle 1,0,1\rangle$ is tangent to the curve at P. Since $|\underline{r}'(0)| = \sqrt{2}$, the two unit tangent vectors are $\pm(1/\sqrt{2})\langle 1,0,1\rangle$.

25. By (15.9), $\int_0^{\pi/4} (\sin t\,\underline{i} - \cos t\,\underline{j} + \tan t\,\underline{k})dt = \int_0^{\pi/4} \sin t\,dt\,\underline{i} - \int_0^{\pi/4} \cos t\,dt\,\underline{j} + \int_0^{\pi/4} \tan t\,dt\,\underline{k} = -\cos t\Big|_0^{\pi/4}\underline{i} - \sin t\Big|_0^{\pi/4} + \ln|\sec t|\Big|_0^{\pi/4}\underline{k}$
$= (-1/\sqrt{2} + 1)\underline{i} - (1/\sqrt{2})\underline{j} + \ln\sqrt{2}\,\underline{k}$.

28. $\underline{r}'(t) = 2\underline{i} - 4t^3\underline{j} + 6t^{1/2}\underline{k} \Longrightarrow \underline{r}(t) = 2t\,\underline{i} - t^4\underline{j} + 4t^{3/2}\underline{k} + \underline{c}$, where $\underline{c}$ is a constant vector. To find $\underline{c}$, note that $\underline{r}(0) = \underline{c}$ and that we desire $\underline{r}(0) = \underline{i} + 5\underline{j} + 3\underline{k}$. Thus $\underline{c} = \underline{i} + 5\underline{j} + 3\underline{k}$, and $\underline{r}(t) = (2t+1)\underline{i} + (5-t^4)\underline{j} + (4t^{3/2}+3)\underline{k}$.

31. From #19, $\langle 1,1,0\rangle$ is tangent to C at P(1,0,4) and is, thus, the normal vector for the normal plane. Its equation is, by (14.30), $1(x-1) + 1(y-0) + 0(z-4) = 0$ or $x + y - 1 = 0$.

37 and 40. These proofs follow from (15.6), writing out the components and differentiating them as in Chapter 3.

43. Same as above except that (15.9) is used. For example:
(b) If $\underline{u} = f\underline{i} + g\underline{j} + h\underline{k}$, $c\underline{u} = cf\underline{i} + cg\underline{j} + ch\underline{k}$ and

$$\int c\underline{u} = \left(\int cf\right)\underline{i} + \left(\int cg\right)\underline{j} + \left(\int ch\right)\underline{k} \quad \text{(By 15.9))}$$

$$= c\left(\int f\right)\underline{i} + c\left(\int g\right)\underline{j} + c\left(\int h\right)\underline{k} \quad \text{(Property of integrals)}$$

$$= c\left[\left(\int f\right)\underline{i} + \left(\int g\right)\underline{j} + \left(\int h\right)\underline{k}\right] = c\int \underline{u} \quad \text{(By 15.9))}.$$

46. $\underline{u}\cdot\underline{v} = 2te^{-t} - 6te^{-t} + t^2 = t^2 - 4te^{-t}$ and so $D_t(\underline{u}\cdot\underline{v}) = 2t + 4te^{-t} - 4e^{-t}$ directly. The reader should verify that the same answer is obtained if (15.8(iii)) is used, i.e., $D_t(\underline{u}\cdot\underline{v}) = \underline{u}\cdot\underline{v}' + \underline{u}'\cdot\underline{v}$.
$\underline{u} \times \underline{v} = (6t + t^2e^{-t})\underline{i} + (t^2e^{-t}-2t)\underline{j} - 8te^{-t}\underline{k} \Longrightarrow D_t(\underline{u} \times \underline{v}) = (6 + (2t-t^2)e^{-t})\underline{i} + ((2t-t^2)e^{-t} - 2)\underline{j} + (8t-8)e^{-t}\underline{k}$. (Again verify by using (15.8(iv).)

EXERCISES 15.3, page 734

1. Position: $\underline{r}(t) = 2t\underline{i} + (4t^2+1)\underline{j}$, $\underline{r}(1) = 2\underline{i} + 5\underline{j}$.
Velocity: $\underline{r}'(t) = 2\underline{i} + 8t\underline{j}$, $\underline{r}'(1) = 2\underline{i} + 8\underline{j}$.
Acceleration: $\underline{r}''(t) = 8\underline{j}$, $\underline{r}''(1) = 8\underline{j}$.
Speed: $|\underline{r}'(t)| = \sqrt{4+64t^2} = 2\sqrt{1+16t^2}$. The path is the parabola $y = x^2+1$.

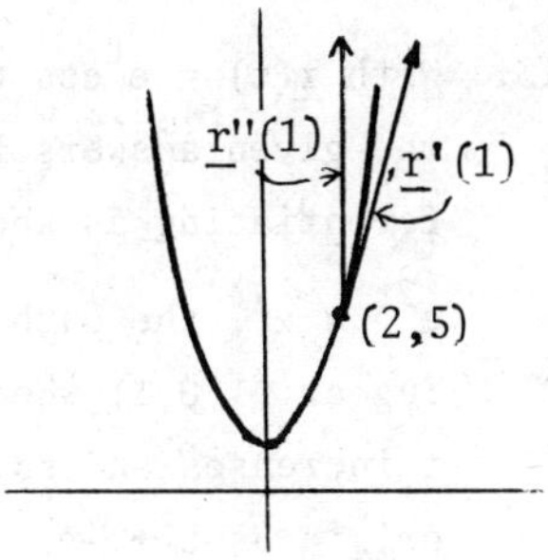

4. With $\underline{r}(t) = t^{1/2}\underline{i} + (1+t^{1/2})\underline{j}$, $\underline{r}(4) = 2\underline{i} + 3\underline{j}$;
$\underline{r}'(t) = \frac{1}{2}t^{-1/2}(\underline{i} + \underline{j})$, $\underline{r}'(4) = \frac{1}{4}(\underline{i} + \underline{j})$;
$\underline{r}''(t) = -\frac{1}{4}t^{-3/2}(\underline{i} + \underline{j})$, $\underline{r}''(t) = -\frac{1}{32}(\underline{i} + \underline{j})$;
$|\underline{r}'(t)| = \frac{1}{2}t^{-1/2}\sqrt{2} = \frac{1}{\sqrt{2}}t^{-1/2}$. The path is the right half of the line $y = 1 + x$.

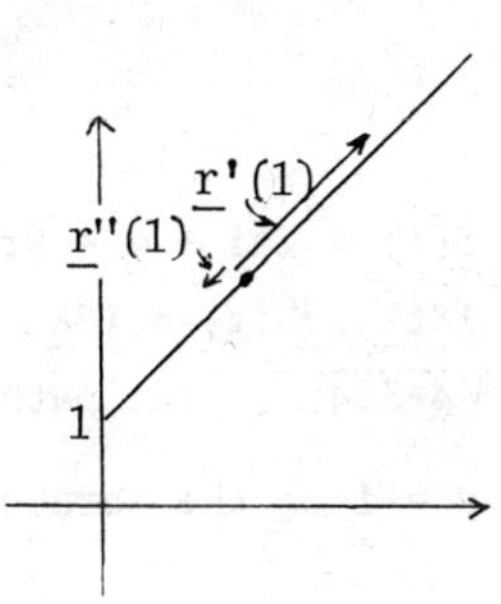

7. $\underline{r}(t) = e^{2t}\underline{i} + e^{-t}\underline{j}$, $\underline{r}(0) = \underline{i} + \underline{j}$; $\underline{r}'(t) = 2e^{2t}\underline{i} - e^{-t}\underline{j}$, $\underline{r}'(0) = 2\underline{i} - \underline{j}$; $\underline{r}''(t) = 4e^{2t}\underline{i} + e^{-t}\underline{j}$, $\underline{r}''(0) = 4\underline{i} + \underline{j}$; $|\underline{r}'(t)| = \sqrt{e^{4t} + e^{-2t}}$. The path is the curve $y = 1/\sqrt{x}$.

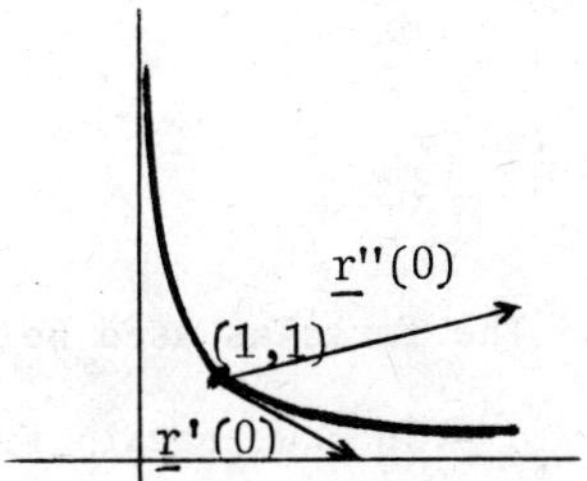

10. $\underline{r}(t) = t^2\underline{i} + t^3\underline{j} + t\underline{k}$; $\underline{r}'(t) = 2t\underline{i} + 3t^3\underline{j} + \underline{k}$; $\underline{r}''(t) = 2\underline{i} + 6t\underline{j} + 0\underline{k}$; $|\underline{r}'(t)| = \sqrt{4t^2 + 9t^4 + 1}$. (Vectors are not shown in the sketch; only the path is which is above the curve $y = x^{3/2}$.)

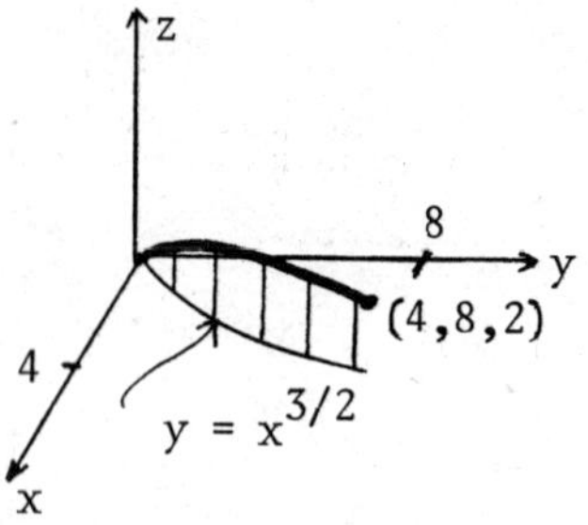

13. With $\underline{r}(t) = e^t\cos t\ \underline{i} + e^t\sin t\ \underline{j} + e^t\underline{k}$, the given answers follow by simply differentiating as above. Since $x^2+y^2 = e^{2t} = z^2$, the path is on this cone starting at (1,0,1) when t = 0, moving up as t increases and spiraling in toward the origin as $t \to -\infty$.

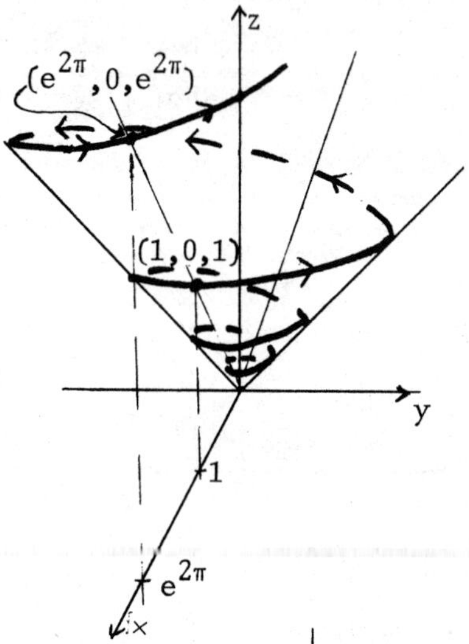

16. $\underline{r}(t) = 2t\underline{i} + \underline{j} + 9t^2\underline{k}$; $\underline{r}'(t) = 2\underline{i} + 18t\underline{k}$; $\underline{r}''(t) = 18\underline{k}$, $|\underline{r}'(t)| = \sqrt{4+324t^2}$. The path is in the plane y = 1 on the parabola $z = \frac{9}{4}x^2$.

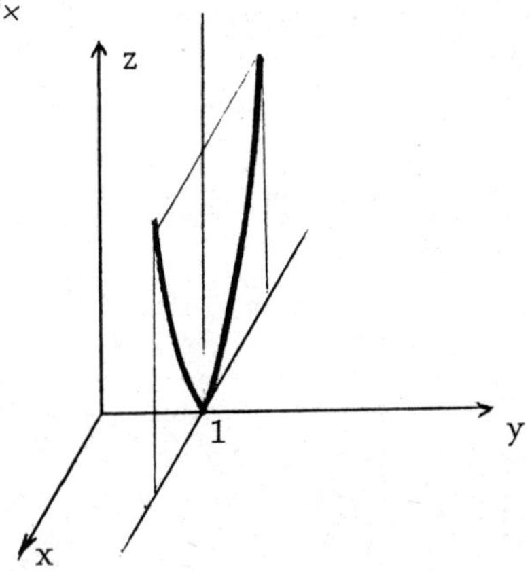

19. The formulas used below are found on p. 734 of the text. With $v_0 = 1500$, $\alpha = 30^\circ$, we obtain from (15.13), $\underline{r}(t) = (1500\cos 30^\circ)\underline{i} + (-\frac{1}{2}gt^2 + (1500\sin 30^\circ)t)\underline{j} = 750\sqrt{3}\ t\ \underline{i} + (-\frac{1}{2}gt^2 + 750t)\underline{j}$.

(a) $\underline{r}'(t) = 750\sqrt{3}\ \underline{i} + (-gt + 750)\underline{j}$.

(b) Maximum altitude $= (v_0^2\sin^2\alpha)/2g \approx (1500^2(1/2)^2)/64 \approx 8{,}789$ ft.

(c) Range $= (v_0^2\sin 2\alpha)/g = (1500^2\sin 60^\circ)/g \approx 1500^2(\sqrt{3}/2)32 \approx 60{,}890$ ft.

(d) The ground is struck when y becomes 0. Setting y (the coefficient of $\underline{j}$ in $\underline{r}(t)$ equal to 0, we obtain $-\frac{1}{2}gt^2 + 750t = 0 \Rightarrow t = 0$ or $t = 1500/g$. The positive solution is desired at which, from (a), the velocity is $\underline{r}'(1500/g) = 750\sqrt{3}\,\underline{i} - 750\,\underline{j}$, and the speed then is $|\underline{r}'(1500/g)| = \sqrt{(750\sqrt{3})^2 + 750^2} = 750\sqrt{3+1} = 1500$ ft/sec.

22. We start with $\underline{r}''(t) = -g\,\underline{j}$ and initial conditions $\underline{r}'(0) = 1800\,\underline{i}$, $\underline{r}(0) = 1000\,\underline{j}$ corresponding to the projectile being fired horizontally in the positive x direction from a height of 1000 ft. $\underline{r}''(t) = -g\,\underline{j} \Rightarrow \underline{r}'(t) = -gt\,\underline{j} + \underline{c}$. Setting $t = 0$ we get $\underline{c} = \underline{r}'(0) = 1800\,\underline{i}$ so that $\underline{r}'(t) = 1800\underline{i} - gt\,\underline{j}$. From this we get $\underline{r}(t) = 1800t\,\underline{i} - \frac{1}{2}gt^2\underline{j} + \underline{d}$. Setting $t = 0$: $\underline{d} = \underline{r}(0) = 1000\,\underline{j}$ and, finally, $\underline{r}(t) = 1800t\,\underline{i} + (1000 - \frac{1}{2}gt^2)\underline{j}$. The ground is hit when $y = 1000 - \frac{1}{2}gt^2 = 0 \Rightarrow t^2 = \frac{2000}{g} \approx \frac{2000}{32} \Rightarrow t \approx \sqrt{\frac{1000}{16}} \approx 7.9$ sec. At that instant, $x \approx 1800(7.9) = 19{,}220$ ft.

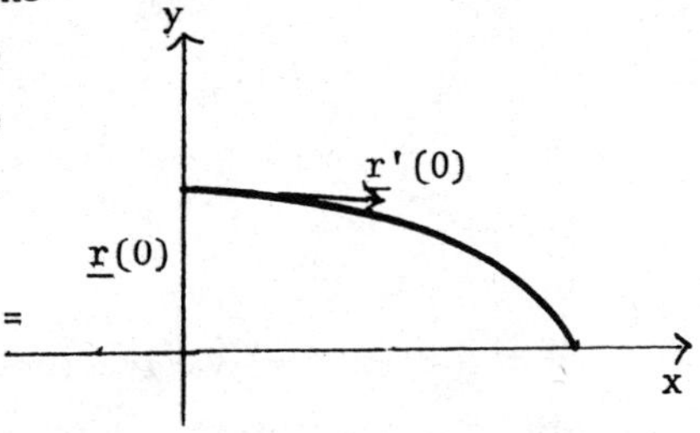

EXERCISES 15.4, page 740

1. When $x = 1$, $y' = -3$, $y'' = -6$ and $K = |-6|/(1+9)^{3/2} = 6/10^{3/2}$.

4. When $x = 2$, $y' = 1$, $y'' = -1$ and $K = |-1|/(1+1)^{3/2} = 1/2^{3/2}$.

7. Eliminating the parameter, the curve has equation $y = \sqrt{x+1}$. Thus $y' = 1/2\sqrt{x+1}$, $y'' = -1/4\sqrt{x+1}^3$. $x = 3 \Rightarrow y' = 1/4$, $y'' = -1/32$. Thus $K = \dfrac{|-1/32|}{(1+1/16)^{3/2}} = \dfrac{1/32}{(17/16)^{3/2}} = \dfrac{64}{(17\sqrt{17})32} = \dfrac{2}{17\sqrt{17}}$.

10. The rectangular equation is $y = \sqrt{x^2-9}$, and $y' = x/\sqrt{x^2-9}$, $y'' = (\sqrt{x^2-9} - x^2/\sqrt{x^2-9})/(x^2-9)$. When $x = 5$, $y' = 5/4$ and $y'' = (4-25/4)/16 = -9/64$. Thus $K = \dfrac{9/64}{(1+25/16)^{3/2}} = \dfrac{9}{41^{3/2}}$.

13. Using the method of implicit differentiation, $9x^2 + 4y^2 = 36 \Rightarrow 18x + 8yy' = 0 \Rightarrow y' = -9x/4y \Rightarrow y'' = -\dfrac{9}{4}\dfrac{y-xy'}{y^2}$. Substituting for y' and using the original equation, $y'' = -\dfrac{9}{4}\dfrac{y-x(-9x/4y)}{y^2} = -\dfrac{9}{4}\dfrac{4y^2+9x^2}{4y^3} = -\dfrac{9}{4}\cdot\dfrac{36}{4y^3} = -\dfrac{81}{4y^3}$.

Thus $K = \dfrac{(81/4|y|^3)}{(1+81x^2/16y^2)^{3/2}} = \dfrac{81}{4|y|^3}\;\dfrac{64|y|^3}{(16y^2+81x^2)^{3/2}} = \dfrac{81(16)}{(16y^2+81x^2)^{3/2}}$. From the original equation, $81x^2 = 9(9x^2) = 9(36-4y^2) = 324 - 36y^2$, $-3 \le y \le 3$. Thus $K = 81(16)(16y^2+324-36y^2)^{-3/2} = 1296(324-20y^2)^{-3/2}$, $-3 \le y \le 3$. We seek the maximum value of K on the interval [-3,3]. Now, $K' = 1296(3/2)(324-20y^2)^{-5/2} \cdot (40y) = 0$ at $y = 0$. By the first derivative test K attains a local minimum at $y = 0$ since $K' < 0$ if $y < 0$, $K' > 0$ if $y > 0$. Thus K attains its maximum value at an end point. Since K is an even function of y, we have $K(-3) = K(3) =$ maximum value. When $y = \pm 3$ then $x = 0$, so this maximum curvature occurs at the points $(0,\pm 3)$.

16. $y' = \cos x$, $y'' = -\sin x \Rightarrow K = |\sin x|/(1 + \cos^2 x)^{3/2}$. Consider the interval $0 \le x \le \pi$ where $\sin x \ge 0$. Then $K = \sin x(1 + \cos^2 x)^{-3/2} \Rightarrow K' = \sin x(-\frac{3}{2})(1 + \cos^2 x)^{-5/2}(-2 \cos x \sin x) + \cos x(1 + \cos^2 x)^{-3/2} = \frac{\cos x}{(1 + \cos^2 x)^{5/2}}(3 \sin^2 x + 1 + \cos^2 x) = 0$ only when $\cos x = 0 \Rightarrow x = \pi/2$. By the first derivative test K attains its maximum there. By periodicity of sin x, K attains this maximum at every 0 of cos x, namely at $x = \pi/2 + n\pi$, where n is an integer.

19. From (15.19), $K = 0$ if $y'' = 0$. $y = \sinh x \Rightarrow y' = \cosh x \Rightarrow y'' = \sinh x = 0$ only at $x = 0$.

EXERCISES 15.5, page 744

NOTE: Exercises 1 and 9, 4 and 12, 7 and 15 are done together. The abbreviations TC and NC are used for the tangential and normal components of acceleration, respectively.

1,9. $\underline{r}' = 2t\underline{i} + 3\underline{j} \Rightarrow \underline{r}'' = 2\underline{i}$, $|\underline{r}'| = \sqrt{4t^2+9}$, and $TC = \frac{\underline{r}'\cdot\underline{r}''}{|\underline{r}'|} = \frac{4t}{\sqrt{4t^2+9}}$. Next,

$$\underline{r}' \times \underline{r}'' = \begin{vmatrix} \underline{i} & \underline{j} & \underline{k} \\ 2t & 3 & 0 \\ 2 & 0 & 0 \end{vmatrix} = -6\underline{k} \text{ and } NC = \frac{|\underline{r}' \times \underline{r}''|}{|\underline{r}'|} = \frac{6}{\sqrt{4t^2+9}}. \text{ Finally}$$

$$K = \frac{NC}{|\underline{r}'|^2} = \frac{6}{(4t^2+9)^{3/2}}.$$

4,12. $\underline{r}' = 4\underline{i} + 2t\underline{j} + 4t\underline{k} \Rightarrow \underline{r}'' = 2\underline{j} + 4\underline{k}$, $|\underline{r}'| = \sqrt{16+4t^2+16t^2} = 2\sqrt{4+5t^2}$ and $TC = \frac{10t}{\sqrt{4+5t^2}}$. Next, $\underline{r}' \times \underline{r}'' =$

$$\begin{vmatrix} \underline{i} & \underline{j} & \underline{k} \\ 4 & 2t & 4t \\ 0 & 2 & 4 \end{vmatrix} = 0\underline{i} - 16\underline{j} + 8\underline{k} \text{ and } NC = \frac{\sqrt{16^2+8^2}}{2\sqrt{4+5t^2}} = \frac{4\sqrt{5}}{\sqrt{4+5t^2}}, \; K = \frac{\sqrt{5}}{(4+5t^2)^{3/2}}.$$

7,15. $\underline{r}' = -4 \sin t\, \underline{i} + 9 \cos t\, \underline{j} + \underline{k} \Rightarrow \underline{r}'' = -4 \cos t\, \underline{i} - 9 \sin t\, \underline{j}$, $|\underline{r}'| = \sqrt{16 \sin^2 t + 81 \cos^2 t + 1}$ and $TC = \frac{16 \sin t \cos t - 81 \sin t \cos t}{|\underline{r}'|} =$

$\frac{-65 \sin t \cos t}{\sqrt{16 \sin^2 t + 81 \cos^2 t + 1}}$. Next, $\underline{r}' \times \underline{r}'' = 9 \sin t \, \underline{i} + 4 \cos t \, \underline{j} + 36 \, \underline{k}$ so that NC $= \sqrt{81 \sin^2 t + 16 \cos^2 t + 36^2}/|\underline{r}'|$ and $K = \sqrt{81 \sin^2 t + 16 \cos^2 t + 36^2}/|\underline{r}'|^3$ ($|\underline{r}'|$ as above.)

17. Hint: For motion on $y = x^2$ if $\underline{r} = f(t)\underline{i} + g(t)\underline{j}$ then $g(t) = f(t)^2$ and the horizontal component of velocity is $f'(t)$. Given $f'(t) = 3$, it follows that $f(t) = 3t+c$. The time at which P(1,1) is obtained determines c (e.g. if $t = 1$ is chosen, then $c = -2$). This gives $f(t)$ and $g(t)$ from which $\underline{r}'$, $\underline{r}''$, etc. can be determined.

19. If $|\underline{r}'(t)| = c$ for all t then $|\underline{r}'(t)|^2 = c^2$ or $\underline{r}'(t) \cdot \underline{r}'(t) = c^2$. Now, differentiate this equation using (15.9(iii)).

22. Let $\underline{r}(t) = a \cos t \, \underline{i} + b \sin t \, \underline{j} + ct \, \underline{k}$. Then $\underline{r}'(t) = -a \sin t \, \underline{i} + b \cos t \, \underline{j} + c \, \underline{k}$, $|\underline{r}'(t)| = \sqrt{a^2 \sin^2 t + b^2 \cos^2 t + c^2}$, $\underline{r}''(t) = -a \cos t \, \underline{i} - b \sin t \, \underline{j} + 0 \, \underline{k}$. Then

$$\underline{r}'(t) \times \underline{r}''(t) = \begin{vmatrix} \underline{i} & \underline{j} & \underline{k} \\ -a \sin t & b \cos t & c \\ -a \cos t & -b \sin t & 0 \end{vmatrix} = bc \sin t \, \underline{i} - ac \cos t \, \underline{j} +$$

$(ab \sin^2 t + ab \cos^2 t)\underline{k} = bc \sin t \, \underline{i} - ac \cos t \, \underline{j} + ab \, \underline{k}$, and $|\underline{r}'(t) \times \underline{r}''(t)| = \sqrt{b^2c^2 \sin^2 t + a^2c^2 \cos^2 t + a^2b^2}$. By (15.25), $K = \frac{|\underline{r}'(t) \times \underline{r}''(t)|}{|\underline{r}'(t)|^3}$

$$= \frac{(b^2c^2 \sin^2 t + a^2c^2 \cos^2 t + a^2b^2)^{1/2}}{(a^2 \sin^2 t + b^2 \cos^2 t + c^2)^{3/2}}.$$

25. Let $\underline{r}(t) = 2 \sin t \, \underline{i} + 3 \cos t \, \underline{j}$. The point $P(1, 3\sqrt{3}/2)$ occurs when $t = \pi/6$. $\underline{r}'(t) = 2 \cos t \, \underline{i} - 3 \sin t \, \underline{j} \Rightarrow \underline{r}'(\pi/6) = \sqrt{3} \, \underline{i} - \frac{3}{2}\underline{j} \Rightarrow |\underline{r}'(\pi/6)| = \sqrt{3+9/4} = \sqrt{21}/2$. $\underline{r}''(t) = -2 \sin t \, \underline{i} - 3 \cos t \, \underline{j} \Rightarrow \underline{r}''(\pi/6) = -\underline{i} - \frac{3\sqrt{3}}{2} \underline{j}$.

$$\text{Thus } \underline{r}'(\pi/6) \times \underline{r}''(\pi/6) = \begin{vmatrix} i & j & k \\ \sqrt{3} & -3/2 & 0 \\ -1 & -3\sqrt{3}/2 & 0 \end{vmatrix} = (-\frac{9}{2} - \frac{3}{2})\underline{k} = -6\underline{k} \Rightarrow$$

$$|\underline{r}'(\pi/6) \times \underline{r}''(\pi/6)| = 6 \text{ and } K = \frac{6}{(\sqrt{21}/2)^3} = \frac{48}{21\sqrt{21}}.$$

EXERCISES 15.7 (Review), page 752

1. The path is the portion of the parabola $y = 4x - x^2$ lying to the right of the y-axis. The answers for $\underline{r}'$ and $\underline{r}''$ are found by simple differentiation.

4. (a) $\underline{r}' = 3\underline{i} + 3t^2\underline{j} + 4t^3\underline{k}$, $\underline{r}'' = 6t\underline{j} + 12t^2\underline{k}$,
(c) $|\underline{r}'(t)| = \sqrt{34}$.

7. Integrating, $\underline{u}(t) = (-e^{-t} + a)\underline{i} + (2\cos 2t + b)\underline{j} + (2t^{3/2} + c)\underline{k}$. Thus $\underline{u}(0) = (-1+a)\underline{i} + (2+b)\underline{j} + c\underline{k}$ and we must choose $a = b = c = 0$ to get $\underline{u}(0) = -\underline{i} + 2\underline{j} + 0\underline{k}$.

10. At $x = 0$, $y' = 1$, $y'' = 2$ and $K = 2/2^{3/2} = 1/\sqrt{2}$.

13. $\underline{r}' = 2\cos 2t\,\underline{i} - \sin t\,\underline{j} \Longrightarrow \underline{r}'' = -4\sin 2t\,\underline{i} - \cos t\,\underline{j}$ and $|\underline{r}'| = \sqrt{4\cos^2 2t + \sin^2 t}$ and $TC = \underline{r}'\cdot\underline{r}''/|\underline{r}'| = \dfrac{-8\sin 2t\cos 2t + \sin t\cos t}{\sqrt{4\cos^2 2t + \sin^2 t}}$.

Next, $\underline{r}' \times \underline{r}'' = -(2\cos 2t\cos t + 4\sin 2t\sin t)\underline{k}$ and $NC = \dfrac{|\underline{r}' \times \underline{r}''|}{|\underline{r}'|} = \dfrac{|2\cos 2t\cos t + 4\sin 2t\sin t|}{\sqrt{4\cos^2 2t + \sin^2 t}}$.

CHAPTER 16
PARTIAL DIFFERENTIATION

EXERCISES 16.1, page 760

1. As a polynomial, there are no restrictions on x or y, and the domain is the entire plane. $f(-2,5) = 2(-2)-5^2 = -4-25 = -29$. $f(5,-2) = 2(5)-(-2)^2 = 10-4 = 6$. $f(0,-2) = 2(0)-(-2)^2 = 0-4 = -4$.

4. The form of f(r,s) requires $1-r \geq 0$ and $s \neq 0$. Thus the domain is $\{(r,2)\colon r \leq 1,\ s \neq 0\}$. $f(1,1) = \sqrt{1-1} - e^{1/1} = 0-e = -e$. $f(0,4) = \sqrt{1-0} - e^{0/4} = 1-1 = 0$. $f(-3,3) = \sqrt{1-(-3)} - e^{-3/3} = \sqrt{4} - e^{-1} = 2-e^{-1}$.

7. The presence of tan v means that the values $v = \frac{\pi}{2} + n\pi$, n any integer, must be ruled out. Also, ln p requires $p > 0$. Thus the domain is $\{(r,s,v,p)\colon v \neq \frac{\pi}{2} + n\pi,\ p > 0\}$. $f(3,-1,\frac{\pi}{4},e) = 3(-1)^2\tan\frac{\pi}{4} + 4(-1)(\frac{\pi}{4})\ln e = 3 - \pi(1) = 3-\pi$.

10. $\dfrac{f(x+h,y) - f(x,y)}{h} = \dfrac{[y^2+3(x+h)] - [y^2+3x]}{h} = \dfrac{3h}{h} = 3.$

$\dfrac{f(x,y+h) - f(x,y)}{h} = \dfrac{[(y+h)^2 + 3x] - [y^2+3x]}{h} = \dfrac{2yh+h^2}{h} = 2y + h.$

13. The graph of f is the graph of the equation $z = f(x,y) = \sqrt{1-x^2-y^2}$. Since $z \geq 0$ and $z^2 = 1-x^2-y^2$, or $x^2+y^2+z^2 = 1$, the graph is the upper half of the sphere of radius 1 centered at the origin.

16. $z = \sqrt{x^2+y^2} \Longrightarrow z \geq 0$ and $z^2 = x^2+y^2$. Thus, the graph is the upper half of the cone $x^2+y^2-z^2 = 0$.

19. The level curves of f(x,y) are graphs of the equations $y^2-x^2 = c$; hyperbolas with asymptotes $y = \pm x$, opening up about the y-axis if $c > 0$, the x-axis if $c < 0$; the 2 lines $y = \pm x$ if $c = 0$.

22. The level curves of $f(x,y) = xy$ are the familiar hyperbolas $xy = c$ or $y = c/x$, if $c \neq 0$, or the 2 coordinate axes of $c = 0$.

25. The level surfaces of $f(x,y,z) = x^2+y^2+z^2$ are graphs of $x^2+y^2+z^2 = c$, spheres of radius $\sqrt{c}$ (if $c \geq 0$) centered at the origin.

28. The level surfaces of $f(x,y,z) = x^2+y^2-z^2$ are graphs of $x^2+y^2-z^2 = c$, hyperboloids of 1 sheet if $c > 0$, hyperboloids of 2 sheets if $c < 0$, a cone if $c = 0$. (Compare 14.46-14.48.)

EXERCISES 16.2, page 767

1. As a rational function, $(x^2-2)/(3+xy)$ is continuous at (0,0) since the denominator is nonzero there. Thus $\lim\limits_{(x,y)\to(0,0)} \dfrac{x^2-2}{3+xy} = \dfrac{0-2}{3+0} = -\dfrac{2}{3}$.

4. If (x,y) approaches $(0,0)$ along the x-axis, then $y = 0$, and $f(x,y) = f(x,0) = x^2/3x^2 = 1/3$, and the limit attained along this path is 1/3. If, however, (x,y) approaches $(0,0)$ along the y-axis, then $x = 0$, and $f(x,y) = f(0,y) = 5y^2/4y^2 = 5/4$, and the limit attained along this path is 5/4. Since different paths produce different limits, $\lim_{(x,y)\to(0,0)} \frac{x^2-2xy+5y^2}{3x^2+4y^2}$ does not exist.

7. We can write $f(x,y) = \frac{x(y-2)-(y-2)}{(x^2-2x+1)+(y^2-4y+4)} = \frac{(x-1)(y-2)}{(x-1)^2+(y-2)^2}$. If $(x,y) \to (1,2)$ along the horizontal line $y = 2$, then $y-2 = 0$ and $f(x,2) = 0/(x-1)^2 = 0$. (If $(x,y) \to (1,2)$ along the vertical line $x = 1$, the limit is also 0, but no conclusion may be drawn from this.) If $(x,y) \to (1,2)$ along the line with slope m through $(1,2)$, then $y-2 = m(x-1)$ and $f(x,y) = \frac{m(x-1)^2}{(x+1)^2+m^2(x-1)^2} = \frac{m}{1+m^2}$. A different limit is attained for each m!. Since different paths produce different limits, $\lim_{(x,y)\to(1,2)} f(x,y)$ does not exist.

10. If $(x,y,z) \to (0,0,0)$ on a coordinate axis, the limit approached is 0 (e.g. on the x-axis, $y = z = 0$ and $f(x,0,0) = 0/x^2 = 0 \to 0$). However, if the approach is on the line $x = t$, $y = t$, $z = t$, then $f(x,y,z) = f(t,t,t) = \frac{3t^2}{3t^2} = 1 \to 1$. Since different paths to the origin produce different limiting values, the limit itself does not exist.

13. As a quotient of polynomials, f is continuous wherever the denominator is non-zero, i.e. off the cone $x^2+y^2-z^2 = 0$.

16. As the composite of the continuous functions $g(t) = \sqrt{t}$ and $h(x,y) = 25-x^2-y^2$, f has domain, $\{(x,y): 25-x^2-y^2 \geq 0\} = \{(x,y): x^2+y^2 \leq 25\}$, the set of points on and within the circle $x^2+y^2 = 25$. f is continuous on its domain by the definition of continuity and the remarks under Figure 16.11 on page 765.

19. $g(z) = z^2+1 \Longrightarrow g(f(x,y)) = f(x,y)^2+1 = (x + \tan y)^2 + 1$ for all points (x,y) except those on the horizontal lines $y = \pi/2 + n\pi$, n an integer, where tan y is not defined.

22. $g(t) = t^2 \Longrightarrow g(f(x,y,z)) = (f(x,y,z))^2 = (2x+ye^z)^2$.

25.

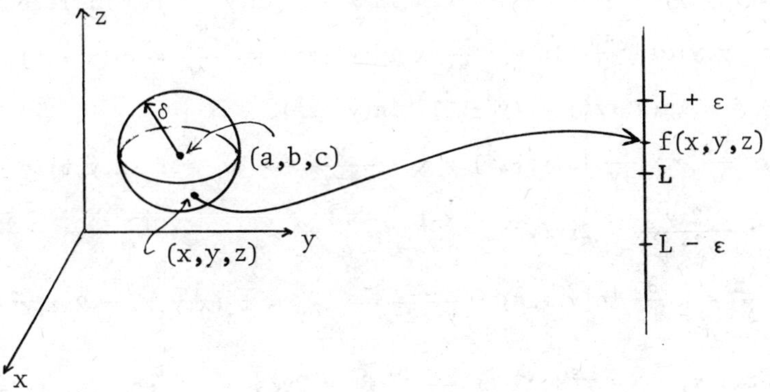

28. From the definitions of continuity of f at (a,b) and limits, we know that for any $\varepsilon > 0$, there is a $\delta > 0$ so that $|f(x,y)-f(a,b)| < \varepsilon$ whenever $\sqrt{(x-a)^2 + (y-b)^2} < \delta$. Let us choose $\varepsilon = f(a,b) > 0$. Then if (x,y) is within the corresponding δ units of (a,b) we have $|f(x,y)-f(a,b)| < f(a,b)$ which is equivalent to $-f(a,b) < f(x,y) - f(a,b) < f(a,b)$ or $0 < f(x,y) < 2f(a,b)$ as desired.

EXERCISES 16.3, page 772

NOTE: Remember that partial derivatives are computed by treating all variables, other than the one you are differentiating, as constants.

1. $f_x(x,y) = \frac{\partial}{\partial x}(2x^4y^3) - \frac{\partial}{\partial x}(xy^2) + \frac{\partial}{\partial x}(3y+1) = 8x^3y^3 - y^2 + 0.$

$f_y(x,y) = \frac{\partial}{\partial y}(2x^4y^3) - \frac{\partial}{\partial y}(xy^2) + \frac{\partial}{\partial y}(3y+1) = 6x^4y^2 - 2xy + 3.$

4. $f_s(s,t) = \frac{\partial}{\partial s}(\frac{t}{s}) - \frac{\partial}{\partial s}(\frac{s}{t}) = -\frac{t}{s^2} - \frac{1}{t}$.

$f_t(s,t) = \frac{\partial}{\partial t}(\frac{t}{s}) - \frac{\partial}{\partial t}(\frac{s}{t}) = \frac{1}{s} + \frac{s}{t^2}$.

7. Using properties of the ln function, we may write $f(t,v) = \frac{1}{2}\ln\frac{t+v}{t-v} = \frac{1}{2}\ln(t+v) - \frac{1}{2}\ln(t-v)$. Thus $f_t(t,v) = \frac{1}{2(t+v)} - \frac{1}{2(t-v)}$; $f_u(t,v) = \frac{1}{2(t+v)} - \frac{(-1)}{2(t-v)}$ which reduce to the answers given.

10. $f_x(x,y) = \sqrt{4x^2-y^2}\,\frac{\partial}{\partial x}\sec x + (\frac{\partial}{\partial x}\sqrt{4x^2-y^2})\sec x = \sqrt{4x^2-y^2}\sec x\tan x + 4x\sec x/\sqrt{4x^2-y^2}$.

$f_y(x,y) = \sqrt{4x^2-y^2}\,\frac{\partial}{\partial y}\sec x + (\frac{\partial}{\partial y}\sqrt{4x^2-y^2})\sec x = 0 + (-y/\sqrt{4x^2-y^2})\sec x = -y\sec x/\sqrt{4x^2-y^2}$.

13. Let $w = f(x,y,z) = (y^2+z^2)^x$. Then $\ln w = x \ln(y^2+z^2)$. Differentiating with respect to x yields $\frac{\partial}{\partial x} \ln w = \frac{\partial}{\partial x} x \ln(y^2+z^2) \Rightarrow \frac{w_x}{w} = \ln(y^2+z^2) \Rightarrow w_x = f_x(x,y,z) = w \ln(y^2+z^2) = (y^2+z^2)^x \ln(y^2+z^2)$.

Similarly, $\frac{w_y}{w} = x \frac{\partial}{\partial y} \ln(y^2+z^2) = x \cdot \frac{2y}{y^2+z^2} \Rightarrow w_y = f_y(x,y,z) = w \cdot \frac{2xy}{y^2+z^2} = (y^2+z^2)^x \cdot \frac{2xy}{(y^2+z^2)} = 2xy(y^2+z^2)^{x-1}$.

Finally, $\frac{w_z}{w} = x \frac{\partial}{\partial z} \ln(y^2+z^2) = \frac{2xz}{y^2+z^2} \Rightarrow w_z = f_z(x,y,z) = 2xz(y^2+z^2)^{x-1}$.

16. $f_r = 3r^2 \tan s$, $f_s = r^3 \sec^2 s + \frac{e^{v^2}}{2\sqrt{s}}$, $f_v = 2v\sqrt{s}\, e^{v^2} - \cos 2p$, $f_p = 2v \sin 2p$.

19. $w_x = y^4 - 4xy^3 + 8x \Rightarrow w_{xy} = 4y^3 - 12xy^2$.

$w_y = 4xy^3 - 6x^2y^2 - 3 \Rightarrow w_{yx} = 4y^3 - 12xy^2 = w_{xy}$.

22. $w_x = 2xy^2e^{x^2} = 2/x^3y^3 \Rightarrow w_{xy} = 4xye^{x^2} + 6/x^3y^4$.

$w_y = 2ye^{x^2} - 3/x^2y^4 \Rightarrow w_{yx} = 2y(2x)e^{x^2} + 6/x^3y^4 = w_{xy}$.

25. $w_x = 6xy^3z + 2y^4z^2 - 0 \Rightarrow w_{xy} = \frac{\partial}{\partial y} w_x = 18xy^2z + 8y^3z^2 \Rightarrow w_{xyz} = \frac{\partial}{\partial z} w_{xy} = 18xy^2 + 16y^3z$.

28. $v_z = \dfrac{y \frac{\partial}{\partial z}(x^2+z^4)}{(x^2+z^4)} = \dfrac{4yz^3}{x^2+z^4} \Rightarrow v_{zz} = \dfrac{(x^2+z^4)\partial(4yz^3)/\partial z - 4yz^3\partial(x^2+z^4)/\partial z}{(x^2+z^4)^2}$

$= \dfrac{(x^2+z^4)(12yz^2) - 4yz^3(4z^3)}{(x^2+z^4)^2} = \dfrac{4z^2y(3x^2+3z^4-4z^4)}{(x^2+z^4)^2} = \dfrac{4z^2(3x^2-z^4)}{(x^2+z^4)^2}\, y$.

Thus $v_{zzy} = \dfrac{4z^2(3x^2-z^4)}{(x^2+z^4)^2}$.

31. $w_r = 4r^3s^3t - 3ts^2e^{rt} \Rightarrow w_{rr} = 12r^2s^3t - 3t^2s^2e^{rt}$ and $w_{rs} = 12r^3s^2t - 6tse^{rt}$. Thus $w_{rrs} = w_{rsr} = 36r^2s^2t - 6t^2se^{rt}$. Next, $w_s = 3r^4s^2t - 6se^{rt} \Rightarrow w_{sr} = 12r^3s^2t - 6tse^{rt} \Rightarrow w_{srr}$ is the same as w_{rrs}.

34. $f_x = \dfrac{\frac{\partial}{\partial x}(y/x)}{1+(y/x)^2} = \dfrac{-y/x^2}{(x^2+y^2)/x^2} = \dfrac{-y}{x^2+y^2} \Rightarrow f_{xx} = \dfrac{2xy}{(x^2+y^2)^2}$.

$f_y = \dfrac{\frac{\partial}{\partial y}(y/x)}{1+(y/x)^2} = \dfrac{1/x}{(x^2+y^2)/x^2} = \dfrac{x}{x^2+y^2} \Rightarrow f_{yy} = \dfrac{-2xy}{(x^2+y^2)^2}$.

Thus $f_{xx} + f_{yy} = 0$.

37. $\frac{\partial w}{\partial x} = -\sin(x-y) + \frac{1}{x+y} \Longrightarrow \frac{\partial^2 w}{\partial x^2} = -\cos(x-y) - \frac{1}{(x+y)^2}$.

$\frac{\partial w}{\partial y} = -\sin(x-y)(-1) + \frac{1}{x+y} = \sin(x-y) + \frac{1}{x+y} \Longrightarrow \frac{\partial^2 w}{\partial y^2} = \cos(x-y)(-1) - \frac{1}{(x+y)^2} =$

$\frac{\partial^2 w}{\partial x^2}$. Thus $\frac{\partial^2 w}{\partial x^2} - \frac{\partial^2 w}{\partial y^2} = 0$.

40. For each partial derivative it is convenient to solve PV = cn T for the desired dependent variable. $V = \frac{cn\ T}{P} \Longrightarrow \frac{\partial V}{\partial T} = \frac{cn}{P}$. $T = \frac{PV}{cn} \Longrightarrow \frac{\partial T}{\partial P} = \frac{V}{cn}$.

$P = \frac{cn\ T}{V} \Longrightarrow \frac{\partial P}{\partial V} = -\frac{cn\ t}{V^2}$. Thus $\frac{\partial V}{\partial T} \cdot \frac{\partial T}{\partial P} \cdot \frac{\partial P}{\partial V} = \frac{cn}{P} \cdot \frac{V}{cn} \cdot (-\frac{cn\ T}{V^2}) = -\frac{cn\ T}{PV}$.

But, from PV = cn T, it follows that $\frac{cn\ T}{PV} = 1$ which gives the desired result.

43. The rates of change of V in the directions of the x-, y-, z-axes are simply V_x, V_y, V_z, respectively. Thus we compute $V_x = -200x/(x^2+y^2+z^2)^2$, $V_y = -200y/(x^2+y^2+z^2)^2$, $V_z = -200z/(x^2+y^2+z^2)^2$.

(a) $V_x(2,-1,1) = -400/(4+1+1)^2 = -400/36$;

(b) $V_y(2,-1,1) = 200/36$; (c) $V_z(2,-1,1) = -200/36$.

EXERCISES 16.4, page 780

1. We seek η_1 and η_2 so that $\Delta w = f(x+\Delta x, y+\Delta y) - f(x,y) = f_x(x,y)\Delta x + f_y(x,y) + \eta_1\Delta x + \eta_2\Delta y$. With $w = f(x,y) = 4y^2 - 3xy + 2x$ we have:

$$\begin{aligned}\Delta w &= [4(y+\Delta y)^2 - 3(x+\Delta x)(y+\Delta y) + 2(x+\Delta x)] - [4y^2 - 3xy + 2x] \\ &= 8y\Delta y + 4\Delta y^2 - 3x\Delta y - 3y\Delta x - 3\Delta x\Delta y + 2\Delta x \\ &= (-3y+2)\Delta x + (8y-3x)\Delta y - (3\Delta y)\Delta x + (4\Delta y)\Delta y.\end{aligned}$$

Noticing that the coefficients of Δx and Δy in the 1st two expressions are f_x and f_y, we get the desired form by choosing $\eta_1 = -3\Delta y$ and $\eta_2 = 4\Delta y$.

4. $$\begin{aligned}\Delta w &= [2(x+\Delta x)^2 - (x+\Delta x)(y+\Delta y)^2 + 3(y+\Delta y)] - [2x^2 - xy^2 + 3y] \\ &= 4x\Delta x + 2\Delta x^2 - (2xy\Delta y + x\Delta y^2 + y^2\Delta x + 2y\Delta x\Delta y + \Delta x\Delta y^2) + 3\Delta y \\ &= (4x-y^2)\Delta x + (-2xy+3)\Delta y + (2\Delta x + 2y\Delta y)\Delta x + (-x\Delta y - \Delta x\Delta y)\Delta y\end{aligned}$$

and we get the desired form by choosing $\eta_1 = 2\Delta x + 2y\Delta y$ and $\eta_2 = -x\Delta y - \Delta x\Delta y$. (There are other choices too.)

7. $dw = \frac{\partial w}{\partial x}dx + \frac{\partial w}{\partial y}dy$ yields the given answer immediately.

10. $$\begin{aligned}dw &= \frac{\partial}{\partial x}(\ln(x^2+y^2) + x\tan^{-1}y)dx + \frac{\partial}{\partial y}(\ln(x^2+y^2) + x\tan^{-1}y)dy \\ &= (\frac{2x}{x^2+y^2} + \tan^{-1}y)dx + (\frac{2y}{x^2+y^2} + \frac{x}{1+y^2})dy.\end{aligned}$$

13. First, $\frac{\partial w}{\partial x} = \frac{(x+y+z)\frac{\partial}{\partial x}(xyz) - (xyz)\frac{\partial}{\partial x}(x+y+z)}{(x+y+z)^2} = \frac{(x+y+z)(yz) - xyz}{(x+y+z)^2}$

$= \frac{yz(y+z)}{(x+y+z)^2}$

and $\frac{\partial w}{\partial y}$, $\frac{\partial w}{\partial z}$ are computed similarly. Then

$dw = \frac{\partial w}{\partial x}dx + \frac{\partial w}{\partial y}dy + \frac{\partial w}{\partial z}dx = \frac{yz(y+z)dx + xz(x+z)dy + xy(x+y)dz}{(x+y+z)^2}$.

16. $dw = \frac{\partial w}{\partial x}dx + \frac{\partial w}{\partial y}dy + \frac{\partial w}{\partial z}dz + \frac{\partial w}{\partial t}dt + \frac{\partial w}{\partial v}dv = (2xy^3zt^{-1}v^4)dx + (3x^2y^2zt^{-1}v^4)dy +$
$(x^2y^3t^{-1}v^4)dz + (-x^2y^3zt^{-2}v^4)dt + (4x^2y^3zt^{-1}v^3)dv$.

19. With $x = f(x,y,z) = x^2z^3 - 3yz^2 + x^{-3} + 2y^{1/2}z$, we want to estimate $\Delta w =$ $f(1.02,3.97,1.96) - f(1,4,2)$. With $x = 1$, $x + \Delta x = 1.02$ we get $\Delta x = .02$, and similarly $\Delta y = -.03$, $\Delta z = -.04$, and $\Delta w \approx dw = f_x(1,4,2)(.02) +$ $f_y(1,4,2)(-.03) + f_z(1,4,2)(-.04)$. Now

$f_x(x,y,z) = 2xz^3 - 3x^{-4}$, $f_x(1,4,2) = 2 \cdot 8 - 3 = 13$,

$f_y(x,y,z) = -3z^2 + y^{-1/2}z$, $f_y(1,4,2) = -3 \cdot 4 + 4^{-1/2} \cdot 2 = -11$

$f_z(x,y,z) = 3x^2z^2 - 6yz + 2y^{1/2}$, $f_z(1,4,2) = 12 - 48 + 4 = -32$

and so $\Delta w \approx 13(.02) - 11(-.03) - 32(-.04) = .26 + .33 + 1.28 = 1.87$.

22. Let $w = f(x,y) = x^{2/5}/y^4$ so that $f(32.03,1.95)$ is the number desired. We can estimate this by using $\Delta w = f(32.03,1.95) - f(32,2) \approx dw = f_x(32,2)(.03) +$ $f_y(32,2)(-.05)$. ($\Delta x = .03$ since $x = 32$, $x+\Delta x = 32.03$; $\Delta y = -0.05$ since $y = 2$, $y+\Delta y = 1.95$). Now, $f(32,2) = 32^{2/5}/2^4 = 2^2/16 = 1/4$. ($32^{1/5} = 2$)

$f_x(x,y) = 2x^{-3/5}/5y^4$, $f_x(32,2) = 2(2^{-3})/5(2^4) = 1/320$.

$f_y(x,y) = -4x^{2/5}/y^5$, $f_y(32,2) = -4(2^2)/2^5 = -1/2$.

Thus $f(32,03,1.95) \approx \frac{1}{4} + \frac{.03}{320} + \frac{.05}{2} \approx .250 + .001 + .025 = 0.276$.

25. Thinking of the given dimensions as interior and recalling that the volume of a circular cylinder with base radius r and height h is $V = \pi r^2 h = f(r,h)$, then the volume of tin is $\Delta V = f(1.5+.015, 4+.015) - f(1.5,4)$. Now, $\Delta V \approx dV$ $= f_r\Delta r + f_h\Delta h = (2\pi rh)\Delta r + (\pi r^2)\Delta h$. Evaluating this when $r = 1.5$, $h = 4$, $\Delta r = \Delta h = .015$, we obtain $\Delta V \approx 12\pi(.015) + 2.25\pi(.015) = \frac{57\pi}{4}(.015) \approx .067$ cu. inches.

28. First, we use the data to find k. $PV = kT \Longrightarrow (.5)(64) = k(80) \Longrightarrow k =$ $32/80 = 0.4$. Next, express P in terms of V and T: $P = .4T/V$. We want ΔP if $T = 80$, $T + \Delta T = 76$ (so $\Delta T = -4$) and $V = 64$, $V + \Delta V = 70$ (so $\Delta V = 6$).

$\Delta P \approx dP = P_T\Delta T + P_V\Delta V = \frac{.4}{V}\Delta T - \frac{.4T}{V^2}\Delta V$

$\Delta P \approx \frac{.4}{64}(-4) - \frac{.4}{64^2}(80)(6) = -(\frac{.1}{4} + \frac{1.5}{32}) = -\frac{2.3}{32} \approx -0.072.$

31. For a circular cylinder of base radius r and height h, the surface area is S = (curved surface area) + (area of top) + (area of bottom) = $2\pi rh + 2\pi r^2$. Now, ΔS is the actual error, and $\Delta S \approx dS = S_r\Delta r + S_h\Delta h = (2\pi h + 4\pi r)\Delta r + 2\pi r\Delta h$ where Δr and Δh are the errors in r and h. Substituting $h = 8$, $r = 3$, $\Delta r = \Delta h = \pm.05$, we get
$\Delta S \approx (16\pi+12\pi)(\pm.05) + (6\pi)(\pm.05) = 34\pi(\pm.05) = \pm1.7\pi \approx \pm5.34$ sq. in.

EXERCISES 16.5, page 787

1. $\frac{\partial w}{\partial x} = \frac{\partial w}{\partial u}\frac{\partial u}{\partial x} + \frac{\partial w}{\partial v}\frac{\partial v}{\partial x} = (2u \sin v)(3x^2) + (u^2\cos v)(y^2)$

$\frac{\partial w}{\partial y} = \frac{\partial w}{\partial u}\frac{\partial u}{\partial y} + \frac{\partial w}{\partial v}\frac{\partial v}{\partial y} = (2u \sin v)(-6y) + (u^2\cos v)(2xy).$

4. $\frac{\partial w}{\partial r} = \frac{\partial w}{\partial t}\frac{\partial t}{\partial r} + \frac{\partial w}{\partial v}\frac{\partial v}{\partial r} = (\frac{e^{t/v}}{v})(2r) + (\frac{-te^{t/v}}{v^2})(3r^2)$

$\frac{\partial w}{\partial s} = \frac{\partial w}{\partial t}\frac{\partial t}{\partial s} + \frac{\partial w}{\partial v}\frac{\partial v}{\partial s} = (\frac{e^{t/v}}{v})(-2s) + (\frac{-te^{t/v}}{v^2})(3s^2).$

7. First note that $\frac{\partial}{\partial u}\ln ut = \frac{t}{ut} = \frac{1}{u}$, $\frac{\partial}{\partial t}\ln ut = \frac{u}{ut} = \frac{1}{t}$.

$\frac{\partial r}{\partial u} = \frac{\partial r}{\partial x}\frac{\partial x}{\partial u} + \frac{\partial r}{\partial y}\frac{\partial y}{\partial u} = (3x^2-y^2)(1) + (3-2xy)(-1/u).$

Similarly the other 2 given answers following using, of course, v and t in the formula instead of u.

10. $\frac{\partial s}{\partial y} = \frac{\partial s}{\partial t}\frac{\partial t}{\partial y} + \frac{\partial s}{\partial r}\frac{\partial r}{\partial y} + \frac{\partial s}{\partial u}\frac{\partial u}{\partial y} + \frac{\partial s}{\partial v}\frac{\partial v}{\partial y} = (r^2\tan^{-1}uv)(2yz) + (2tr\tan^{-1}uv)(3z) + (\frac{\partial s}{\partial u})(0) + (\frac{tr^2u}{1+u^2v^2})(xz).$

13. $\frac{\partial w}{\partial t} = \frac{\partial w}{\partial r}\frac{\partial r}{\partial t} + \frac{\partial w}{\partial s}\frac{\partial s}{\partial t} + \frac{\partial w}{\partial v}\frac{\partial v}{\partial t}$

$= (2r)(2 \sin t \cos t) + (-\tan v)(-\sin t) + (-s \sec^2 v)(4).$

16. Here $F(x,y) = x^4 + 2x^2y^2 - 3xy^3 + 2x$ and, using (16.20), $y' = -(4x^3 + 4xy^2 - 3y^3 + 2)/(4x^2y - 9xy^2)$.

19. With $F(x,y,z)$ the left side of the given equation we have $F_x = 2z^3 + 2xy^2$, $F_y = -3z^2 + 2x^2y$, $F_z = 6xz^2 - 6yz + 4$, and the given answers now follow from (16.21).

22. With $F(x,y,z) = yx^2 + z^2 + \cos xyz - 4$, the equation is of the form $F(x,y,z) = 0$ and thus

$$\frac{\partial z}{\partial x} = \frac{F_x}{F_z} = -\frac{2xy - yz \sin xyz}{2z - xy \sin xyz}, \quad \frac{\partial z}{\partial y} = -\frac{F_y}{F_z} = -\frac{x^2 - xz \sin xyz}{2z - xy \sin xyz}.$$

25. Hint: Start with $\frac{\partial w}{\partial r} = \frac{\partial w}{\partial x}\frac{\partial x}{\partial r} + \frac{\partial w}{\partial y}\frac{\partial y}{\partial r} = \frac{\partial w}{\partial x}\cos\theta + \frac{\partial w}{\partial y}\sin\theta$. Then $\frac{\partial^2 w}{\partial r^2} = \frac{\partial}{\partial x}\left(\frac{\partial w}{\partial r}\right)\frac{\partial x}{\partial r} + \frac{\partial}{\partial y}\left(\frac{\partial w}{\partial r}\right)\frac{\partial y}{\partial r} = \left(\frac{\partial^2 w}{\partial x^2}\cos\theta + \frac{\partial^2 w}{\partial x\partial y}\sin\theta\right)\cos\theta + \left(\frac{\partial^2 w}{\partial y\partial x}\cos\theta + \frac{\partial^2 w}{\partial y^2}\sin\theta\right)\sin\theta$. Similarly, compute $\frac{\partial w}{\partial \theta}$ and $\frac{\partial^2 w}{\partial \theta^2}$, simplify--assuming equality of the mixed derivative--and verify that the desired identity is satisfied.

28. $w = f(u)$, $u = x^2+y^2 \Longrightarrow \frac{\partial w}{\partial x} = f'(u)\frac{\partial u}{\partial x} = f'(u)(2x)$, $\frac{\partial w}{\partial y} = f'(u)\frac{\partial u}{\partial y} = f'(u)(2y)$.

Then $y\frac{\partial w}{\partial x} - x\frac{\partial w}{\partial y} = f'(u)(2xy) - f'(u)(2xy) = 0$.

29. Hint: $\frac{\partial^2 w}{\partial x^2} = \frac{\partial}{\partial x}\left(\frac{\partial w}{\partial x}\right) = \frac{\partial}{\partial x}\left(\frac{\partial w}{\partial u}\frac{\partial u}{\partial x} + \frac{\partial w}{\partial v}\frac{\partial v}{\partial x}\right)$. Now use the sum, product and chain rules, e.g.

$$\frac{\partial}{\partial x}\left(\frac{\partial w}{\partial u}\frac{\partial u}{\partial x}\right) = \frac{\partial w}{\partial u}\frac{\partial^2 u}{\partial x^2} + \frac{\partial u}{\partial x}\frac{\partial}{\partial x}\left(\frac{\partial w}{\partial u}\right) \quad \text{(product rule)}$$

and

$$\frac{\partial}{\partial x}\left(\frac{\partial w}{\partial u}\right) = \frac{\partial}{\partial u}\left(\frac{\partial w}{\partial u}\right)\frac{\partial u}{\partial x} + \frac{\partial}{\partial v}\left(\frac{\partial w}{\partial u}\right)\frac{\partial v}{\partial x} = \frac{\partial^2 w}{\partial u^2}\frac{\partial u}{\partial x} + \frac{\partial^2 w}{\partial v\partial u}\frac{\partial v}{\partial x} \quad \text{(chain rule)}.$$

31. Differentiate each side with respect to R_i. On the left side we get $\frac{\partial}{\partial R_i}(R^{-1}) = -R^{-2}\frac{\partial R}{\partial R_i}$. On the right side, since $R_1, R_2, \ldots, R_n$ are independent variables, we get $\frac{\partial}{\partial R_i}\left(\frac{1}{R_1} + \frac{1}{R_2} + \ldots + \frac{1}{R_i} + \ldots + \frac{1}{R_n}\right) = 0+0+\ldots+\frac{\partial}{\partial R_i}(R_i^{-1}) + \ldots + 0 = -R_i^{-2}$. Thus we obtain $-R^{-2}\frac{\partial R}{\partial R_i} = -R_i^{-2} = -\frac{1}{R_i^2} \Longrightarrow \frac{\partial R}{\partial R_i} = \frac{R^2}{R_i^2}$.

34. If s is the length of the equal sides, θ the included angle, then by elementary trigonometry the height is $s \cos\frac{\theta}{2}$ and the base is $2(s \sin\frac{\theta}{2})$. Thus the area, A, is

$$A = \frac{1}{2}(2s \sin\frac{\theta}{2})(s \cos\frac{\theta}{2}) = \frac{s^2}{2}(2 \sin\frac{\theta}{2}\cos\frac{\theta}{2}) = \frac{s^2}{2}\sin\theta.$$

We are given that $\frac{ds}{dt} = 0.1$ ft/hr and $\frac{d\theta}{dt} = 2^\circ/\text{hr} = \frac{\pi}{90}$ rad/hr. Thus $\frac{dA}{dt} = \frac{\partial A}{\partial s}\frac{ds}{dt}\;\frac{\partial A}{\partial \theta}\frac{d\theta}{dt} = (s \sin\theta)(.1) + \frac{s^2}{2}\cos\theta)(\frac{\pi}{90})$.

(Note: $\pi/90$ must be used rather than 2°. The differentiation formulas for the trig functions were all derived in terms of radian measure, not degree measure!) Finally, when $s = 20$, $\theta = 60^\circ = \pi/3$, $\frac{dA}{dt} = (20\frac{\sqrt{3}}{2})(.1) + (200\cdot\frac{1}{2})\cdot(\frac{\pi}{90}) = \sqrt{3} + \frac{10\pi}{9} \approx 5.22$ sq ft/hr.

EXERCISES 16.6, page 796

1. First, $\underline{u} = \cos\frac{\pi}{4}\,\underline{i} + \sin\frac{\pi}{4}\,\underline{j} = \frac{1}{\sqrt{2}}(\underline{i} + \underline{j})$ is a unit vector in the direction $\theta = \frac{\pi}{4}$. Next computing f_x and f_y we get $\nabla f(x,y) = (2x-5y)\underline{i} + (-5x+6y)\underline{j}$ and $D_{\underline{u}}f(x,y) = \nabla f(x,y)\cdot\underline{u} = \frac{2x-5y-5x+6y}{\sqrt{2}} = \frac{y-3x}{\sqrt{2}}$. At $P(3,-1)$, $D_{\underline{u}}f(3,-1) = \frac{-1-9}{\sqrt{2}} = \frac{-10}{\sqrt{2}}$. (Observe that we could have evaluated $\nabla f(3,-1) = 11\underline{i} - 21\underline{j}$ first and then form $\nabla f(3,-1)\cdot\underline{u}$ to get the same result.)

4. A unit vector parallel to $\underline{a} = -\underline{i} + 4\underline{j}$ is $\underline{u} = \frac{1}{|\underline{a}|}\,\underline{a} = \frac{1}{\sqrt{17}}(-\underline{i} + 4\underline{j})$. Next, $\nabla f(x,y) = (2x \ln y)\underline{i} + \frac{x^2}{y}\,\underline{j}$ so that $\nabla f(5,1) = 25\underline{j}$. Thus $D_{\underline{u}}f(5,1) = \nabla f(5,1)\cdot\underline{u}$ $= \frac{0+4(25)}{\sqrt{17}} = \frac{100}{\sqrt{17}}$.

7. $\underline{a} = \langle 5,1\rangle \Longrightarrow |\underline{a}| = \sqrt{26}$, and a unit vector parallel to $\underline{a}$ is $\underline{u} = \frac{1}{|\underline{a}|}\,\underline{a} = \frac{1}{\sqrt{26}}\langle 5,1\rangle$. Next, $\nabla f(x,y) = \langle \cos^2 y, -2x \cos y \sin y\rangle$, $\nabla f(2,\pi/4) = \langle (1/\sqrt{2})^2, -4(1/\sqrt{2})^2\rangle = \langle 1/2,-2\rangle$. Thus $D_{\underline{u}}f(2,\pi/4) = \nabla f(2,\pi/4)\cdot\underline{u} = \langle 1/2,-2\rangle\cdot \frac{1}{\sqrt{26}}\langle 5,1\rangle = \frac{1}{\sqrt{26}}(\frac{5}{2} - 2) = \frac{1}{2\sqrt{26}}$.

10. $|\underline{a}| = \sqrt{14} \Longrightarrow \underline{u} = \frac{1}{\sqrt{14}}(2\underline{i} - 3\underline{j} + \underline{k})$. $\nabla f(x,y,z) = f_x\underline{i} + f_y\underline{j} + f_z\underline{k} = (2x+4y)\underline{i}$ $+ (3z+4x)\underline{j} + 3y\,\underline{k}$, $\nabla f(1,0,-5) = 2\underline{i} - 11\underline{j} + 0\underline{k}$, and $D_{\underline{u}}f(1,0,-5) = \frac{1}{\sqrt{14}}(4+33+0)$ $= \frac{37}{\sqrt{14}}$.

13. A unit vector parallel to $\underline{a}$ is $\underline{u} = \frac{1}{\sqrt{10}}\langle -3,0,1\rangle$; $\nabla f(x,y,z) = \langle y+z,x+2y+z,x+y\rangle$ so that $\nabla f(5,7,1) = \langle 8,20,12\rangle$ and $D_{\underline{u}}f(5,7,1) = \nabla f(5,7,1)\cdot\underline{u} = \frac{-24+12}{\sqrt{10}} = \frac{-12}{\sqrt{10}}$.

16. $\overrightarrow{PQ} = \pi/3,-\pi/6 = \frac{\pi}{6}\;2,1$. Thus a unit vector parallel to $\overrightarrow{PQ}$ is $\frac{1}{\sqrt{5}}\;2,-1$. Next, $\nabla f(x,y) = \langle 2\cos(2x-y), -\cos(2x-y)\rangle$ so that $\nabla f(-\frac{\pi}{3},\frac{\pi}{6}) =$ $\langle 2\cos(-\frac{5\pi}{6}), -\cos(-\frac{5\pi}{6})\rangle = \langle -\sqrt{3}, \frac{\sqrt{3}}{2}\rangle$. Thus $D_{\underline{u}}f(-\frac{\pi}{3},\frac{\pi}{6}) = \nabla f\cdot\underline{u} = \frac{-2\sqrt{3} - \sqrt{3}/2}{\sqrt{5}}$ $= -\frac{\sqrt{3}}{\sqrt{5}}\cdot\frac{5}{2} = -\frac{\sqrt{3}\sqrt{5}}{2}$. f increases most rapidly in the direction of $\nabla f(-\frac{\pi}{3},\frac{\pi}{6}) = \langle -\sqrt{3}, \sqrt{3}/2\rangle$. (f would decrease most rapidly in the opposite

direction.) The maximum rate of increase of f is $|\nabla f(-\frac{\pi}{3},\frac{\pi}{6})| = \sqrt{3+\frac{3}{4}} = \frac{\sqrt{15}}{2}$.

19. It is given that $T = \frac{k}{\sqrt{x^2+y^2}}$, where k is the constant of proportionality, and that $T(3,4) = 100$. Substituting we get $\frac{k}{\sqrt{9+16}} = \frac{k}{5} = 100$ or $k = 500$. Thus $T = \frac{500}{\sqrt{x^2+y^2}}$, $\nabla T = \left\langle \frac{-500x}{\sqrt{(x^2+y^2)^3}}, \frac{-500y}{\sqrt{(x^2+y^2)^3}} \right\rangle$, so that at P(3,4), $\nabla T(3,4)$ $= \left\langle \frac{-1500}{125}, \frac{-2000}{125} \right\rangle = -\langle 12,16 \rangle$. A unit vector parallel to $\underline{i} + \underline{j} = \langle 1,1 \rangle$ is $\underline{u} = \frac{1}{\sqrt{2}}\langle 1,1 \rangle$, and $D_{\underline{u}}T(3,4) = \frac{-12-16}{\sqrt{2}} = \frac{-28}{\sqrt{2}}$. T increases most rapidly in any direction parallel to $\nabla T(3,4) = -12\underline{i} - 16\underline{j}$. The rate of change of T is 0 in any direction $\underline{a}$ such that $D_{\underline{a}}T(3,4) = \nabla T(3,4)\cdot\underline{a} = 0$. If $\underline{a} = a_1\underline{i} + a_2\underline{j} = \langle a_1,a_2 \rangle$, this results in $-12a_1 - 16a_2 = 0$. One solution is $a_1 = 4$, $a_2 = -3$ and $\underline{a}$ can be any vector of the form $c(4\underline{i} - 3\underline{j})$ where c is any nonzero constant.

22. $\underline{u} = \frac{1}{7}(2\underline{i} + 6\underline{j} - 3\underline{k})$ is a unit vector parallel to $2\underline{i} + 6\underline{j} - 3\underline{k}$. $\nabla T = 8x\underline{i} - 2y\underline{j} + 32z\underline{k}$ so that $\nabla T(4,-2,1) = 32\underline{i} + 4\underline{j} + 32\underline{k}$. Thus $D_{\underline{u}}T(4,-2,1) = \frac{64+24-96}{7} = -\frac{8}{7}$. The direction of most rapid rate of increase of T at (4,-2,1) is $\nabla T(4,-2,1)$ and this maximum rate is $|\nabla T(4,-2,1)|$ $= \sqrt{32^2 + 4^2 + 32^2} = \sqrt{16(129)} = 4\sqrt{129}$.

25. $\nabla(cu) = \langle (cu)_x, (cu)_y \rangle = \langle cu_x, cu_y \rangle = c\langle u_x, u_y \rangle = c\nabla u$.

28. $\nabla(\frac{u}{v}) = \langle (\frac{u}{v})_x, (\frac{u}{v})_y \rangle$. The quotient rule and a rearrangement produces the desired result.

EXERCISES 16.7, page 803

1. Let F(x,y,z) be the left side of the equation rewritten as $4x^2-y^2+3z^2-10 = 0$. Then $\nabla F(x,y,z) = \langle 8x,-2y,6z \rangle$ and at P(2,-3,1), $\nabla F(2,-3,1) = \langle 16,6,6 \rangle$. This is a normal vector for the tangent plane and a direction vector for the normal line. Thus, using P, we get: $16(x-2) + 6(y+3) + 6(z-1) = 0$ for the tangent plane and $\frac{x-2}{16} = \frac{y+3}{6} = \frac{z-1}{6}$ for the normal line.

4. Rewriting as $4x^2-y^2-z = 0$ and with F(x,y,z) being the left side, we have $\nabla F(x,y,z) = \langle 8x,-2y,-1 \rangle$ so that $\nabla F(5,-8,36) = \langle 40,16,1 \rangle$. Thus:

   Tangent plane: $40(x-5) + 16(y+8) - (z-36) = 0$

   Normal line: $\frac{x-5}{40} = \frac{y+8}{16} = \frac{z-36}{-1}$ or $x = 5+40t$, $y = -8+16t$, $z = 36-t$.

7. As above, let $F(x,y,z) = 2e^{-x}\cos y - z$ so that the equation is $F(x,y,z) = 0$. Then $\nabla F(x,y,z) = \langle -2e^{-x}\cos y, -2e^{-x}\sin y, -1\rangle$, $\nabla F(0,\pi/3,1) = \langle -1,-\sqrt{3},-1\rangle$ from which the given answers follow immediately.

10. The equation is $F(x,y,z) = 0$ if $F(x,y,z) = xyz - 4xz^3 + y^3-10$ so that $\nabla F(x,y,z) = \langle yz-4z^3, xz+3y^2, xy-12xz^2\rangle$, $\nabla F(-1,2,1) = \langle -2,11,10\rangle$. Thus

Tangent plane: $-2(x+1) + 11(y-2) + 10(z-1) = 0$

Normal line: $\frac{x+1}{-2} = \frac{y-2}{11} = \frac{z-1}{10}$.

13. As above, with $F(x,y,z) = \frac{x^2}{a^2} - \frac{y^2}{b^2} - \frac{z^2}{c^2} - 1$ we have $\nabla F(x,y,z) = \left\langle \frac{2x}{a^2}, -\frac{2y}{b^2}, -\frac{2z}{c^2}\right\rangle$ and at P_o, $\nabla F(x_o,y_o,z_o) = \left\langle \frac{2x_o}{a^2}, \frac{2y_o}{b}, \frac{2z_o}{c^2}\right\rangle$ and the tangent plane has equation:

$$\frac{2x_o}{a^2}(x-x_o) - \frac{2y_o}{b^2}(y-y_o) - \frac{2z_o}{c^2}(z-z_o) = 0 \text{ or } \frac{xx_o}{a^2} - \frac{yy_o}{b^2} - \frac{zz_o}{c^2} = \frac{x_o^2}{a^2} - \frac{y_o^2}{b^2} - \frac{z_o^2}{c^2}.$$

But the right side is exactly 1 since the coordinates of P_o satisfy the original equation.

16. Let $P_o(x_o,y_o,z_o)$ be a point on the surface so that $x_o^{2/3} + y_o^{2/3} + z_o^{2/3} = a^{2/3}$. With $F(x,y,z) = x^{2/3} + y^{2/3} + z^{2/3} - a^{2/3}$ we have $\nabla F(x,y,z) = \left\langle \frac{2}{3}x^{-1/3}, \frac{2}{3}y^{-1/3}, \frac{2}{3}z^{-1/3}\right\rangle$. Evaluating at P_o we obtain for the tangent plane, after cancelling the common factor of 2/3,

$$x_o^{-1/3}(x-x_o) + y_o^{-1/3}(y-y_o) + z_o^{-1/3}(z-z_o) = 0.$$

Let A,B,C, be the x,y,z intercepts, respectively, of this plane. To find A, set $y = z = 0$ to obtain $x_o^{-1/3}(A-x_o) = y_o^{2/3} + z_o^{2/3} = a^{2/3} - x_o^{2/3}$ which yields $A = x_o+x_o^{1/3}(a^{2/3}-x_o^{2/3}) = x_o^{1/3}a^{2/3}$. Similarly $B = y_o^{1/3}a^{2/3}$, $C = z_o^{1/3}a^{2/3}$ so that $A^2 + B^2 + C^2 = (x_o^{2/3} + y_o^{2/3} + z_o^{2/3})a^{4/3} = a^{2/3}a^{4/3} = a^2$.

19. Direction vectors for the normal lines to the graphs of $F = 0$ and $G = 0$ are ∇F and ∇G, respectively. These lines are orthogonal, then, if and only if, $\nabla F \cdot \nabla G = 0$ or $F_xG_x + F_yG_y + F_zG_z = 0$.

22. The level curves are the lines $3x-2y = c$. To pass through $P(-2,1)$ we must have $3(-2)-2(1) = c$ or $c = -8$. Here $\nabla f = \langle 3,-2\rangle$.

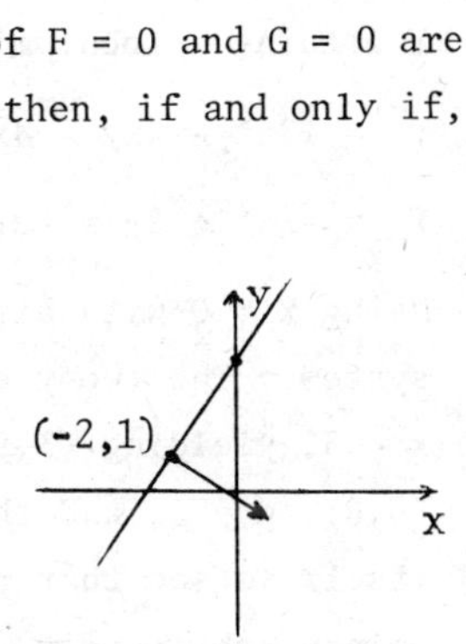

25. The level surfaces are spheres $x^2+y^2+z^2= c$. To pass through $P(-2,1,5)$, c must be 30. $\nabla f(-2,1,5) = \langle -4,2,10\rangle$.

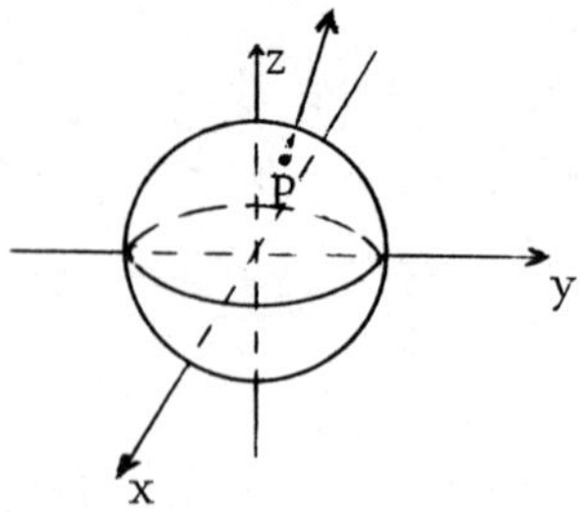

28. The level surfaces are graphs of $x^2+y^2-z^2= c$. To pass through $(3,-1,1)$, c must be 9. Hence the surface is the hyperboloid of one sheet $x^2+y^2-z^2 = 9$ (See #22, Sec. 16.1). $\nabla f(3,-1,1) = \langle 6,-2-2\rangle$.

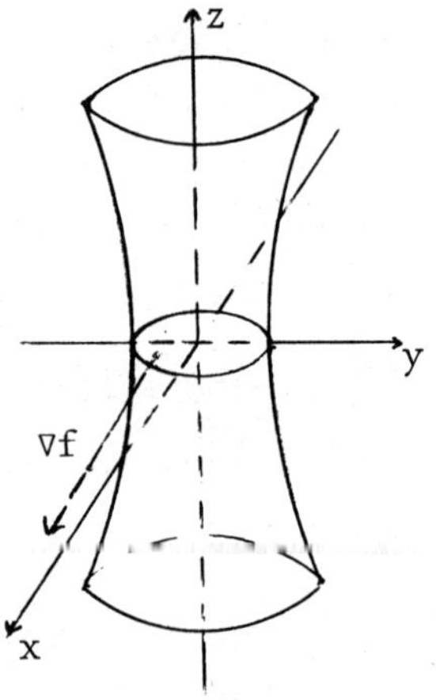

EXERCISES 16.8, page 809

NOTE: In this section $f_{xx}f_{yy} - f_{xy}^2$ will be called g.

1. $\left.\begin{array}{l} f_x = 2x + 2y = 0 \\ f_y = 2x + 6y = 0 \end{array}\right\} \Rightarrow x = y = 0.$ Since $g = 2(6)-4 = 8 > 0$, and $f_{xx} = 2 > 0$,

   f attains a local minimum at $(0,0)$.

4. $f_x = 12x^2 - 4xy = 4x(3x-y) = 0 \Rightarrow x = 0$ or $y = 3x$.

   $f_y = -2x^2 + 2y = -2(x^2 - y) = 0 \Rightarrow y = x^2$.

   Using $x = 0$ with either of the y equations yields $(0,0)$ as a solution of the system. The two y equations yield $3x = x^2$ or $x^2-3x = 0$ with solution $x = 0$, $x = 3$, yielding $(3,9)$ as another point. Now, $g = (24x-4y)(2) - 16x^2$ and at $(0,0)$, $g = 0$, and the test gives no information. However, looking at $f(x,y)$ itself we see that on the x-axis, $y = 0$ and $f(x,0) = x^3$. Thus there are points arbitrarily close to $(0,0)$ at which $f(x,0) > 0$ (if $x > 0$) and $f(x,0) < 0$ (if $x < 0$). Thus there is no local extremum at $(0,0)$. At $(3,9)$,

$g = (72-36)(2) - 16 \cdot 9 = -72 < 0$, and $f_{xx} = 36 > 0$. Thus there is a saddle point at (3,9), no local extremum.

7. $f_x = 4x^3 + 32 = 0 \Rightarrow x^3 = -8 \Rightarrow x = -2$.

$f_y = 3y^2 - 9 = 0 \Rightarrow y^2 = 3 \Rightarrow y = \pm\sqrt{3}$.

$g = (12x^2)(6y) - 0 = 72x^2y$. At $(-2,\sqrt{3})$, $g = 72(4)\sqrt{3} > 0$ and $f_{xx} = 12(4) > 0$. Thus f attains a local minimum at $(-2,\sqrt{3})$. At $(-2,-\sqrt{3})$, $g = -72(4)\sqrt{3} < 0$, and there is no local extremum at $(-2,-\sqrt{3})$.

10. $f_x = \sin y = 0 \Rightarrow y = n\pi$, n an integer.

$f_y = x \cos y = 0 \Rightarrow x = 0$ or $\cos y = 0$.

Since $\cos n\pi = \pm 1 \neq 0$, we must have $x = 0$ to have both equations satisfied. Thus the solutions are $(0,n\pi)$. Now, $f_{xx} = 0$, $f_{xy} = \cos y \Rightarrow g = -\cos^2 y = -1$ whenever $y = n\pi$. Since $g < 0$ at these points, there are no local extrema.

13. $f_x = \cos x + \cos(x+y) = 0$, $f_y = \cos y + \cos(x+y) = 0 \Rightarrow \cos x = \cos y$. If we expand the 2nd equation and use the relations $\sin x = \pm\sqrt{1-\cos^2 x} = \pm\sqrt{1-\cos^2 y}$, $\sin y = \pm\sqrt{1-\cos^2 y}$, we obtain $\cos y + \cos x \cos y - \sin x \sin y = \cos y + \cos^2 y - (\pm\sqrt{1-\cos^2 y})(\pm\sqrt{1-\cos^2 y}) = 0$, or, to simplify things, let $t = \cos y$ to obtain $t+t^2-(\pm\sqrt{1-t^2})(\pm\sqrt{1-t^2}) = 0$. If the radicals have opposite signs, the result is $t+t^2 + (1-t^2) = t+1 = 0$ or $\underline{t = -1}$. If they have the same sign, we get $t+t^2 - (1-t^2) = 2t^2+t-1 = 0$ with solutions $t = 1/2$, $t = -1$. Thus the only solutions are $t = \cos y = -1$ (yielding $y = \pi$), and $\cos y = 1/2$ which yields $y = \pi/3$ and $5\pi/3$. Using these in the $\cos x = \cos y$ relation we get the 5 points to test: $(\pi,\pi), (\frac{\pi}{3},\frac{\pi}{3}), (\frac{5\pi}{3},\frac{5\pi}{3}), (\frac{\pi}{3},\frac{5\pi}{3}), (\frac{5\pi}{3},\frac{\pi}{3})$. But $f_y = 3/2$ at the last 2 points; so only the 1st 3 are solutions of our system. Next, $f_{xx} = -\sin x - \sin(x+y)$, $f_{yy} = -\sin y - \sin(x+y)$, $f_{xy} = -\sin(x+y)$. Thus at (π,π), $g = 0$ and the test fails. However, by noting that $f(\pi,\pi) = 0$ and writing f in the form $f(x,y) = \sin x + \sin y + \sin x \cos y + \cos x \sin y = \sin x(1+\cos y) + \sin y(1+\cos x)$ we see that if $\pi < x, y < 2\pi$, $f(x,y) < 0$ whereas if $0 < x, y < \pi$, $f(x,y) > 0$. Thus there is no extremum at (π,π). At $(\frac{\pi}{3},\frac{\pi}{3})$, $g = (-\sqrt{3})(\sqrt{3}) - (-\sqrt{3}/2)^2 = 9/4 > 0$, and $f_{xx} = -\sqrt{3} < 0$; so f attains a local maximum there. At $(\frac{5\pi}{3},\frac{5\pi}{3})$, $g = (\sqrt{3})(\sqrt{3}) - \sqrt{3}/2)^2 = 9/4 > 0$, and $f_{xx} = \sqrt{3} > 0$; so f attains a local minimum there.

16. $f_x = 2x - 6 \cos y = 0 \Rightarrow x = 3 \cos y$.

$f_y = 6x \sin y = 0 \Rightarrow x = 0$ or $y = n\pi$ (n, integer).

If $x = 0$, then, from the 1st relation, $\cos y = 0$ and $y = \frac{\pi}{2} + n\pi$. If $y = n\pi$, then $x = 3 \cos n\pi = 3(-1)^n$ since $\cos n\pi = 1$ if n is even, -1 if n is odd. Thus there are infinitely many solutions: the points $(0, \frac{\pi}{2} + n\ )$ and $(3(-1)^n, n\pi)$. Computing $f_{xx} = 2$, $f_{yy} = 6x \cos y$, $f_{xy} = 6 \sin y$ we find that $g(0, \frac{\pi}{2} + n\pi) = 2(0) - (6 \sin(\frac{\pi}{2} + n\pi))^2 = -(\pm 6)^2 = -36$. Thus at these points the graph has saddle points. Next, $g(3(-1)^n, n\pi) = 2 \cdot 18(-1)^n \cos n\pi - (\sin n\pi)^2 = 36(-1)^{2n} - 0 = 36 > 0$. Since $f_{xx} > 0$, we have local minima at these points.

19. Let $P(x,y,z)$ be on the graph so that $xy^3z^2 = 16$ and $z^2 = 16/xy^3$. To avoid radicals we'll minimize $d^2(P,0) = f(x,y) = x^2+y^2+z^2 = x^2+y^2+16/xy^3$.

$$f_x = 2x - \frac{16}{x^2y^3} = \frac{2x^3y^3-16}{x^2y^3} = 0 \Rightarrow x^3y^3 = 8 \Rightarrow xy = 2.$$

$$f_y = 2y - \frac{48}{xy^4} = \frac{2xy^5-48}{xy^4} = 0 \Rightarrow xy^5 = 24 \Rightarrow x = 24/y^5.$$

Substituting the last equation into the first yields $\frac{24}{y^5}\,y = 2$ or $y^4 = 12$.

Thus $y = \pm\sqrt[4]{12}$, $x = \frac{24}{\pm 12^{5/4}} = \pm\frac{2}{\sqrt[4]{12}}$, and $z^2 = 16/(\frac{2}{\sqrt[4]{12}})(\sqrt[4]{12})^2 = 8/(\sqrt[4]{12})^2 = 8/\sqrt{12}$ so that $z = \pm\sqrt{8}/\sqrt[4]{12}$. These solutions yield 4 points on the graph since x and y must have the same sign ($xy = 2 > 0$), but z can be of either sign. That all of these points yield a minimum for d^2, and thus for d, is geometrically evident. Also $g = 36$ and $f_{xx} > 0$ produces the result analytically.

22. Let the open box have base dimensions x and y, height z. Then the surface area is $A = xy + 2xz + 2yz$ and the volume is $V = xyz = xy(A-xy)/2(x+y) = (Axy-x^2y^2)/2(x+y)$.

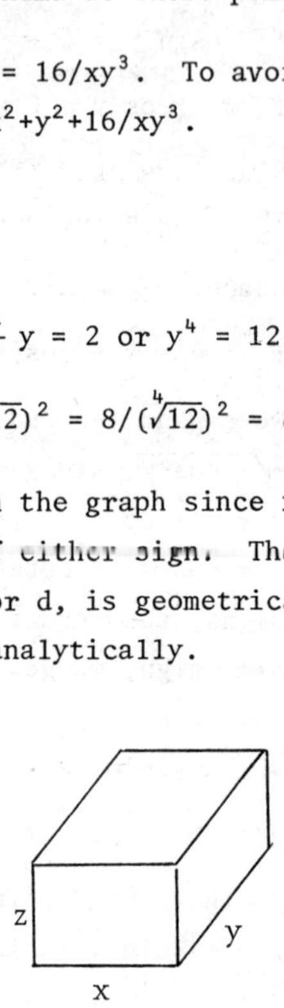

$$V_x = \frac{2(x+y)(Ay-2xy^2)-(Axy-x^2y^2)(2)}{4(x+y)^2} = \frac{y^2(A-x^2-2xy)}{2(x+y)^2}.$$

$$V_y = \frac{2(x+y)(Ax-2x^2y)-(Axy-x^2y^2)(2)}{4(x+y)^2} = \frac{x^2(A-y^2-2xy)}{2(x+y)^2}.$$

In setting these equal to 0, we'll ignore the solution $x = 0$, $y = 0$ as physically meaningless. Thus we get

$$\left.\begin{array}{l} A - x^2 - 2xy = 0 \\ \\ A - y^2 - 2xy = 0 \end{array}\right\} \Rightarrow x^2 - y^2 = 0 \Rightarrow x = y = \sqrt{A/3}.$$

(The value $\sqrt{A/3}$ obtained by setting $y = x$ in either of the 2 equations.)

Thus we must have a square base. The height is $z = \frac{A - xy}{2(x+y)} = \frac{A - A/3}{4\sqrt{A/3}} = \frac{1}{2}\sqrt{\frac{A}{3}}$ which is 1/2 the side of the base.

25. $V = xyz$ where $4x + 3y + z = 12$. Thus $V = xy(12-4x-3y) = 12xy - 4x^2y - 3xy^2$.
$V_x = 12y - 8xy - 3y^2 = y(12-8x-3y) = 0$
$V_y = 12x - 4x^2 - 6xy = 2x(6-2x-3y) = 0$
Ignoring the $x = 0$, $y = 0$ solutions (which produce a minimum of $V = 0$) we get the system $8x+3y = 12$, $2x+3y = 6$ or $6x = 6$. Thus $x = 1$, $y = 4/3$, $z = 4$ are the values obtained and $V = 16/3$. This is a maximum on physical grounds as well as the 2nd derivative test.

28. (See Figure 16.24, p. 810.) The slant side of the triangle is $\frac{x/2}{\cos\theta} = \frac{x\sec\theta}{2}$, and the perimeter of the window is $x + 2y + 2\cdot\frac{x\sec\theta}{2} = 12 \Longrightarrow y = (12 - x - x\sec\theta)/2$. Next, the height of the upper triangle is $\frac{x}{2}\tan\theta$, and the area is $\frac{1}{2}x\cdot\frac{x}{2}\tan\theta = \frac{x^2}{4}\tan\theta$. The total area of the window, then, is $A = xy + \frac{x^2}{4}\tan\theta$. Substituting for y from above, we obtain $A = \frac{x}{2}(12 - x - x\sec\theta) + \frac{x^2}{4}\tan\theta = 6x - \frac{x^2}{4}(2 + 2\sec\theta - \tan\theta)$. First, we compute $A_\theta = -\frac{x^2}{4}(2\sec\theta\tan\theta - \sec^2\theta) = -\frac{x^2\sec\theta}{4}(2\tan\theta - \sec\theta)$. Now, $x^2 > 0$ and $\sec\theta$ is never 0. Thus $A_\theta = 0 \Longrightarrow 2\tan\theta = \sec\theta \Longrightarrow 2\frac{\sin\theta}{\cos\theta} = \frac{1}{\cos\theta} \Longrightarrow \sin\theta = \frac{1}{2} \Longrightarrow \theta = \frac{\pi}{6}$. Next, $A_x = 6 - \frac{x}{2}(2 + 2\sec\theta - \tan\theta)$. Setting $\theta = \frac{\pi}{6}$ and $A_x = 0$, using $\sec\frac{\pi}{6} = 2/\sqrt{3}$ and $\tan\pi/6 = 1/\sqrt{3}$, we obtain $x = \frac{12}{2 + 2\sec\pi/6 - \tan\pi/6} = \frac{12}{2 + 4/\sqrt{3} - 1/\sqrt{3}} = \frac{12\sqrt{3}}{2\sqrt{3} + 3}$. This may be simplified by writing $x = \frac{12\sqrt{3}}{3+2\sqrt{3}}\cdot\frac{3-2\sqrt{3}}{3-2\sqrt{3}} = \frac{36\sqrt{3}-72}{9-12} = 24 - 12\sqrt{3} \approx 3.2/6$. The corresponding value of y is: $y = (1/2)(12 - x - x\sec\theta) = (1/2)(12 - 24 + 12\sqrt{3} - (24 - 12\sqrt{3})(2/\sqrt{3}))$ which reduces to $y = 6-2\sqrt{3} \approx 2.536$.

EXERCISES 16.9, page 816

1. $w = y^2 - 4xy + 4x^2 + \lambda(x^2+y^2-1)$

$$\left.\begin{aligned} w_x &= -4y + 8x + 2\lambda x = -4(y-2x) + 2\lambda x = 0 \\ w_y &= 2y - 4x + 2\lambda y = 2(y-2x) + 2\lambda y = 0 \\ w_\lambda &= x^2 + y^2 - 1 = 0 \end{aligned}\right\}$$

The 1st equation plus twice the 2nd yields $2\lambda(x+2y) = 0 \Longrightarrow \lambda = 0$ or $x = -2y$. If $\lambda = 0$, then from either the 1st or 2nd equation $y = 2x$ which, when substituted into the 3rd, produces $5x^2 = 1$ and the 2 solutions $P_1(\frac{1}{\sqrt{5}}, \frac{2}{\sqrt{5}})$, $P_2(-\frac{1}{\sqrt{5}}, -\frac{2}{\sqrt{5}})$. If $x = -2y$, substitution into the 3rd equation produces

$5y^2 = 1$, and we get 2 more solutions $P_3(\frac{2}{\sqrt{5}}, -\frac{1}{\sqrt{5}})$, $P_4(-\frac{2}{\sqrt{5}}, \frac{1}{\sqrt{5}})$. From the form of $f(x,y) = (y-2x)^2$, we see that f has the minimum value of 0 at P_1 and P_2 (where $y = 2x$) and the maximum value of 5 at P_3 and P_4.

4. $w = x^2 + y^2 + z^2 + \lambda(x+y+z-25)$

$$\left.\begin{aligned} w_x &= 2x + \lambda = 0 \\ w_y &= 2y + \lambda = 0 \\ w_z &= 2z + \lambda = 0 \end{aligned}\right\} \qquad \begin{aligned} \text{Add to get } 2(x+y+z) + 3\lambda &= 0 \\ 2\quad(25) + 3\lambda &= 0 \\ \lambda &= -50/3 \end{aligned}$$

$$w_\lambda = x + y + z - 25 = 0$$

Thus $x = y = z = -\lambda/2 = 25/3$ and f has minimum value 625/3 at this point. (Minimum on geometric grounds since f is the square of the distance of a point from the origin for points on the given plane.)

7. $w = x^2 + y^2 + z^2 + \lambda(x-y-1) + \mu(y^2-z^2-1)$.

$w_x = 2x + \lambda = 0$, $w_y = 2y - \lambda + 2y\mu = 0$, $w_z = 2z - 2z\mu = 0$, $w_\lambda = x-y-1 = 0$, $w_\mu = y^2-z^2-1 = 0$ is our system. From the 3rd equation $2z(1-\mu) = 0 \Longrightarrow z = 0$ or $\mu = 1$. If $z = 0$, then $y = \pm 1$ from the 5th equation which yield, from the 4th equation $x = 2$ and 0. Thus we have the 2 solutions $P_1(2,1,0)$, $P_2(0,-1,0)$. If $\mu = 1$, then the 1st 2 equations become $\lambda = -2x = 4y$ or $x = -2y$. Substituting into the 4th yields $y = -1/3$ but when this value is used in the 5th equation we get $1/9 - z^2 - 1 = 0$ or $z^2 = -8/9$, with no solutions. Thus P_1 and P_2 are the only solutions and f attains a local minimum at each. This is clear from the facts that $f = d^2(0,P)$ and that the plane, $x-y = 1$, and the cylinder, $y^2-z^2 = 1$, intersect in a curve consisting of a right branch (for $y \geq 1$) and a left branch (for $y \leq -1$). P_1 is the point on the right branch closest to the origin ($f(P_1) = 5$), and P_2 is the point on the left branch closest to the origin ($f(P_2) = 1$).

10. $w = (x^2+y^2+z^2+t^2) + \lambda(x+y+2z-1) + \mu(2x-z+t-2) + \nu(y+3z+2y+1)$

$$\begin{aligned} w_x &= 2x + \lambda + 2\mu = 0 \\ w_y &= 2y + \lambda + \nu = 0 \\ w_z &= 2z + 2\lambda - \mu + 3\nu = 0 \\ w_t &= 2t + \mu + 2\nu = 0 \end{aligned}$$

These 4 equations plus the 3 plane equations form a system of 7 equations in 7 unknowns! The solution (details omitted) is $x = 75/68$, $y = 15/68$, $z = -11/68$, $t = -25/68$, $\lambda = -37/34$, $\mu = -19/34$, $\nu = 11/17$. Again on geometric grounds $f(\frac{75}{68}, \frac{15}{68}, -\frac{11}{68}, -\frac{25}{68}) = \frac{6596}{4624}$ is a minimum.

13. Let x and y be the base dimensions and z the height. Then $xyz = 2$. The area of the sides is $2yz + 2xz$; that of the top and bottom is xy each. The cost is c = (cost of sides) + (cost of bottom) + (cost of top) = $(2yz + 2xz) + 2xy + \frac{3}{2}xy = 2yz + 2xz + \frac{7}{2}xy$.

Let $w = 2yz + 2xz + \frac{7}{2}xy + \lambda(xyz - 2)$. Then

$w_x = 2z + \frac{7}{2}y + \lambda yz = 0 \Longrightarrow 2xz + \frac{7}{2}xy + 2\lambda = 0$

$w_y = 2z + \frac{7}{2}x + \lambda xz = 0 \Longrightarrow 2yz + \frac{7}{2}xy + 2\lambda = 0$

$w_z = 2y + 2x + \lambda xy = 0 \Longrightarrow 2yz + 2xz + 2\lambda = 0$

(The equations were multiplied by x, y, z respectively and $xyz = 2$ was used.) Subtracting the 2nd from the 1st we get $2z(x-y) = 0 \Longrightarrow z = 0$ or $y = x$. We reject $z = 0$ on physical grounds. Subracting the 3rd from the 1st we get $y(\frac{7}{2}x - 2z) = 0$ or $z = \frac{7}{4}x$. Using this and $y = x$ in $xyz = 2$ we get $\frac{7}{4}x^3 = 2$ or $x = \sqrt[3]{8/7} = 2/\sqrt[3]{7}$, whence $y = 2/\sqrt[3]{7}$ and $z = 7/2\sqrt[3]{7}$.

16. Hint: For simplicity maximize the square of the area to eliminate radicals. The constraint is $x+y+z = p$.

(19, 22, 25 are 13, 16, 19 of the previous section and will not be included here.)

EXERCISES 16.10 (Review), page 817

1. The domain is $\{(x,y): 36-4x^2+9y^2 \geq 0\}$, the region between the 2 branches of the hyperbola $\frac{x^2}{9} - \frac{y^2}{4} = 1$.

4. Since sec z is undefined when $z = \frac{\pi}{2} + n\pi$, n an integer, the domain is $\{(x,y,z): z \neq \frac{\pi}{2} + n\pi,\ y \neq x\}$.

7. $f_x = \dfrac{(y^2+z^2)(2x)-(x^2+y^2)(0)}{(y^2+z^2)^2} = \dfrac{2x}{y^2+z^2}$

$f_y = \dfrac{(y^2+z^2)(2y)-(x^2+y^2)(2y)}{(y^2+z^2)^2} = \dfrac{(z^2-x^2)(2y)}{(y^2+z^2)^2}$

$f_z = \dfrac{(y^2+z^2)(0)-(x^2+y^2)(2z)}{(y^2+z^2)^2} = -\dfrac{(x^2+y^2)(2z)}{(y^2+z^2)^2}$

10. $f_v = 2v \cos w - w^2 \sin v$, $f_w = -v^2 \sin w + 2w \cos v$.

13. $w_x = -x(x^2+y^2+z^2)^{-3/2} \Longrightarrow w_{xx} = 3x^2(x^2+y^2+z^2)^{-5/2} - (x^2+y^2+z^2)^{-3/2} = (x^2+y^2+z^2)^{-5/2}(2x^2-y^2-z^2)$. Similarly, $w_{yy} = (x^2+y^2+z^2)^{-5/2}(2y^2-x^2-z^2)$ and $w_{zz} = (x^2+y^2+z^2)^{-5/2}(2z^2-x^2-y^2)$. Thus $w_{xx} + w_{yy} + w_{zz} = (x^2+y^2+z^2)^{-5/2}(2x^2-y^2-z^2+2y^2-x^2-z^2+2z^2-x^2-y^2) = 0$.

16. ΔR is the error when $E = 108$, $I = 2$ with errors ΔE and ΔI such that $|\Delta E| \leq .2$, $|\Delta I| \leq .01$. With $R = \frac{E}{I}$, $|\Delta R| \approx |dR| = |R_E \Delta E + R_I \Delta I| = |\frac{1}{I}\Delta E - \frac{E}{I^2}\Delta I| \leq \frac{1}{2}(.2) + \frac{108}{4}(.01) = .1 + .27 = .37$ ohms.

19. The answer is immediate using $\frac{dw}{dt} = \frac{\partial w}{\partial x}\frac{dx}{dt} + \frac{\partial w}{\partial y}\frac{dy}{dt} + \frac{\partial w}{\partial z}\frac{dz}{dt}$.

22. Thinking of C traced out by the position vector $\underline{r}(t) = t\underline{i} + t^2\underline{j} + t^3\underline{k}$, then $\underline{r}'(t) = \underline{i} + 2t\underline{j} + 3t^2\underline{k}$ and at P(2,4,8), which corresponds to t = 2, a tangent vector is $\underline{r}'(2) = \underline{i} + 4\underline{j} + 12\underline{k}$. Since $|\underline{r}'(2)| = \sqrt{161}$, a unit tangent vector at P is $\underline{u} = \frac{1}{\sqrt{161}}(\underline{i} + 4\underline{j} + 12\underline{k})$. Next, with $f(x,y,z) = y^2 + xz$, $\nabla f = z\underline{i} + 2y\underline{j} + x\underline{k}$, $\nabla f(2,4,8) = 8\underline{i} + 8\underline{j} + 2\underline{k}$ and $D_{\underline{u}}f(2,4,8) = \frac{1}{\sqrt{161}}(8+32+24) = \frac{64}{\sqrt{161}}$.

25. With F(x,y) the left side of the equation, $y' = -F_x/F_y = -\frac{(3x^2-4y^3+1)}{(-12xy^2-3)}$.

28. Let the base dimensions be x and y, the height z, the volume V, the cost per sq. in. of sides and top c. Then the area of the top and bottom is xy; the total area of the sides and top is 2yz+2xz+xy. The total cost is, then, C = (cost of bottom)+(cost of sides and top) or C = 2cxy + c(xy+2yz+2xz) = c(3xy+2yz+2xz). Since xyz = V, we get z = V/xy and C = c(3xy+2V/x+2V/y).

$$C_x = c(3y - \frac{2V}{x^2}) = 0 \Longrightarrow 3yx^2 = 2V \quad (x \neq 0)$$

$$C_y = c(3x - \frac{2V}{y^2}) = 0 \Longrightarrow 3xy^2 = 2V \quad (y \neq 0).$$

Subtracting, we get 3xy(x-y) = 0 which implies x = y since $x \neq 0$, $y \neq 0$. Using this relation in either the C_x or C_y equation we obtain $3x^3 = 2V$ or $x = y = \sqrt[3]{\frac{2V}{3}}$. Finally, $z = \frac{V}{xy} = \frac{V}{(2V/3)^{2/3}} = V^{1/3}(\frac{3}{2})^{2/3} = \frac{3}{2}x$. Thus, for minimum cost the box should have a square base and height $1\frac{1}{2}$ times a side of the base.

31. $w = xyz + \lambda(x^2 + 4y^2 + 2z^2 - 8)$

$$w_x = yz + 2x\lambda = 0 \Longrightarrow xyz + 2x^2\lambda = 0$$

$$w_y = xz + 8y\lambda = 0 \Longrightarrow xyz + 8y^2\lambda = 0$$

$$w_z = xy + 4z\lambda = 0 \Longrightarrow xyz + 4z^2\lambda = 0.$$

Adding these last 3 equations and remembering that $x^2+4y^2+2z^2 = 8$, we get $3(xyz) + 2(x^2+4y^2+2z^2)\lambda = 3xyz + 16\lambda = 0$ or $xyz = -16\lambda/3$. Substituting this into these 3 equations gives:

$$\lambda(-\frac{16}{3} + 2x^2) = 0 \Longrightarrow \lambda = 0 \text{ or } x^2 = 8/3$$

$$\lambda(-\frac{16}{3} + 8y^2) = 0 \Longrightarrow \lambda = 0 \text{ or } y^2 = 2/3$$

$$\lambda(-\frac{16}{3} + 4z^2) = 0 \Longrightarrow \lambda = 0 \text{ or } z^2 = 4/3 .$$

If $\lambda = 0$, then in order that all 3 original w_x, w_y, and w_z equations be satisfied, 2 of x, y, and z must be 0 and thus $f(x,y,z) = xyz$ is 0. By checking nearby points, it can be seen that this value is neither a local maximum or minimum. (e.g. if $x = y = 0$, then $z = \pm 2$ and $f(0,0,2) = 0$. But for points on this ellipsoid near (0,0,2) and above the 1st or 3rd quadrant of the xy-plane, $f(x,y,z) > 0$ since $xy > 0$ and z is almost 2; if, however, the point is above the 2nd or 4th quadrant, $f(x,y,z) < 0$ since $xy < 0$ and z is almost 2.) Thus the only solutions are $x = \pm\sqrt{8/3}$, $y = \pm\sqrt{2/3}$, $z = \pm 2/\sqrt{3}$, and we get a total of eight points $(\pm\sqrt{8/3}, \pm\sqrt{2/3}, \pm 2/\sqrt{3})$. The value of f is $8/3\sqrt{3}$ at the 4 points when all coordinates are + or just one is +. These are the local maxima. The value of f is $-8/3\sqrt{3}$ at the other 4 points when all or just 1 coordinate is -. These are the local minima.

34. The volume of the cylinder is $4\pi h$ and that of the cone is $\frac{1}{3}\pi 2^2 k = \frac{4}{3}\pi k$. Thus $4\pi(h + \frac{k}{3}) = 100$ is the volume condition. The curved surface area, S, is = (area of curved cylindrical surface) + (surface area of cone) = $2\pi(2)h + 2\pi\sqrt{k^2+2^2}$. (See #23, Sec. 13.7 for the conical formula.) Thus $S = 4\pi h + 2\pi\sqrt{k^2+4}$. But from the volume constraint $4\pi h = 100 - \frac{4\pi k}{3}$ and, in terms of one variable, $S = 100 - \frac{4\pi k}{3} + 2\pi\sqrt{k^2+4}$.

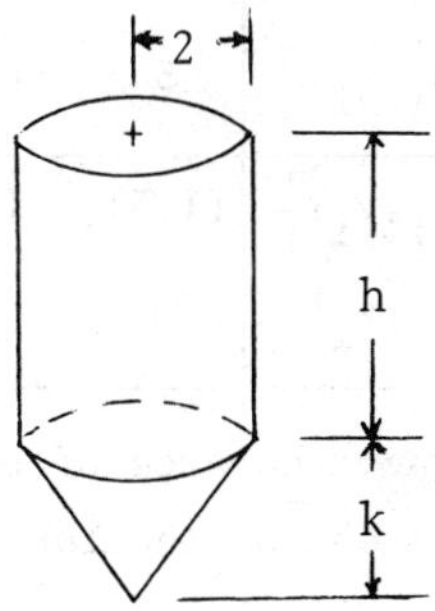

$$\frac{dS}{dk} = -\frac{4\pi}{3} + \frac{2\pi k}{\sqrt{k^2+4}} = 0 \Longrightarrow \frac{2k}{\sqrt{k^2+4}} = \frac{4}{3} \Longrightarrow 6k = 4\sqrt{k^2+4} \Longrightarrow 36k^2 = 16k^2 + 64 \Longrightarrow$$

$k^2 = \frac{16}{5} \Longrightarrow k = 4/\sqrt{5}$, and $h = 100 - 16\pi/3\sqrt{5}$. This is a minimum on physical grounds or by the 2nd derivative test.

CHAPTER 17

MULTIPLE INTEGRALS

NOTE: In this Chapter, as in several previous ones, we will include the integral answers to those odd problems whose complete solutions are not given. This is done in the application sections, Sec. 17.3-17.9.

EXERCISES 17.1, page 825

1. Referring to Figure 17.3, we tabulate our work as follows:

	i (Rectangle #1	1	2	3	4	5	6	7
(a)	(u_i, v_i)	(0,2)	(1,2)	(0,1)	(1,1)	(0,0)	(1,0)	(2,0)
	$f(u_i, v_i)$	5	9	3	7	1	5	9
(b)	(u_i, v_i)	(1,3)	(2,3)	(1,2)	(2,2)	(1,1)	(2,1)	(3,1)
	$f(u_i, v_i)$	11	15	9	13	7	11	15

Since the area of each rectangle is 1, $\Delta A_i = 1$, and the Riemann sum is just the sum of the functional values.

(a) $5+9+3+7+1+5+9 = 39$; (b) $11+15+9+13+7+11+15 = 81$.

4. The sketch illustrates the region, the partition, and the 5 rectangles inside R. Again we tabulate, using

$$f(x,y) = 4 - xy .$$

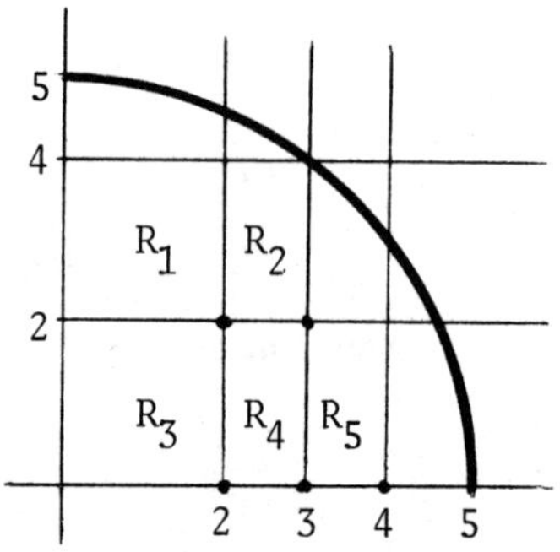

i	1	2	3	4	5
(u_i, v_i)	(2,2)	(3,2)	(2,0)	(3,0)	(4,0)
$f(u_i, v_i)$	0	-2	4	4	4
ΔA_i	4	2	4	2	2
$f(u_i, v_i)\Delta A_i$	0	-4	16	8	8

and the Riemann sum is $0-4+16+8+8 = 28$.

EXERCISES 17.2, page 835

1. $\int_1^2 \int_{-1}^2 (12xy^2 - 8x^3)\,dydx = \int_1^2 [4xy^3 - 8x^3y]_{-1}^{2}\,dx = \int_1^2 [(32x-16x^3) - (-4x+8x^3)]\,dx$

$= \int_1^2 (36x - 24x^3)\,dx = 18x^2 - 6x^4]_1^2 = -36.$

$\int_{-1}^2 \int_1^2 (12xy^2 - 8x^3)\,dxdy = \int_{-1}^2 [6x^2y^2 - 2x^4]_1^2\,dy = \int_{-1}^2 [(24y^2-32) - (6y^2-2)]\,dy$

$= \int_{-1}^2 (18y^2 - 30)\,dy = 6y^3 - 30y]_{-1}^2 = -36.$

4. $\int_{-1}^1 \int_{x^3}^{x+1} (3x + 2y)\,dydx = \int_{-1}^1 [3xy + y^2]_{x^3}^{x+1}\,dx = \int_{-1}^1 [(3x(x+1)+(x+1)^2 -$

$(3x^4+x^6)]\,dx = \int_{-1}^1 (4x^2+5x+1-3x^4-x^6)\,dx = \frac{4}{3}x^3 + \frac{5}{2}x^2 + x - \frac{3}{5}x^5 - \frac{x^7}{7}]_{-1}^1 = \frac{334}{105}.$

7. Recall that $\int e^{ky}dy = \frac{1}{k}e^{ky} + C$. Thus with $k = \frac{1}{x}$ we have,

$\int_1^2 \int_{x^3}^{x} e^{y/x}\,dydx = \int_1^2 xe^{y/x}]_{x^3}^{x}\,dx = \int_1^2 x(e - e^{x^2})\,dx =$

$e\,\frac{x^2}{2} - \frac{1}{2}\,e^{x^2}]_1^2 = (2e - \frac{1}{2}\,e^4) - (\frac{e}{2} - \frac{e}{2}) = \frac{4e - e^4}{2}.$

10. $\int_0^1 \int_y^1 \frac{1}{1+y^2}\,dx\,dy = \int_0^1 \frac{x}{1+y^2}]_y^1\,dy = \int_0^1 (\frac{1}{1+y^2} - \frac{y}{1+y^2})\,dy =$

$\tan^{-1}y - \frac{1}{2}\ln(1+y^2)]_0^1 = \frac{\pi}{4} - \frac{\ln 2}{2}.$

13. We begin by sketching the region R.

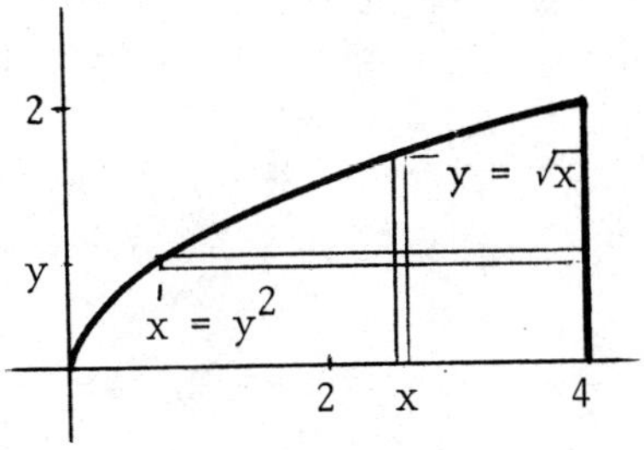

(a) We have to describe R as $R = \{(x,y): a \le x \le b,\ g_1(x) \le y \le g_2(x)\}$ so that for each x in [a,b] we may first integrate f with respect to y, from $g_1(x)$ to $g_2(x)$, and then x from a to b. (The vertical rectangle illustrates this idea.) The x limits are 0 and 4 (0 from the nature of the curve, 4 was given.) For each x in [0,4] y ranges from the x-axis, y = 0 (= $g_1(x)$), to the curve, $y = \sqrt{x}$ (= $g_2(x)$). Thus $\iint_R f\,dA = \int_0^4 \int_0^{\sqrt{x}} f(x,y)\,dy\,dx.$

(b) Now we want to describe R as $R = \{(x,y): c \le y \le d,\ h_1(y) \le x \le h_2(y)\}$ so that for each y in [c,d] we first integrate f with respect to x from $h_1(y)$ to $h_2(y)$, and then y from c to d. (The horizontal rectangle illustrates this.) We see that the y limits are 0 and 2 (0 was given and 2 comes from the upper boundary curve $y = \sqrt{x}$ having maximum value of 2 on $0 \le x \le 4$). For each y in [0,2], x ranges from the curve, now written as $x = y^2$ $(= h_1(y))$, to the line $x = 4$ $(= h_2(y))$. Thus $\iint_R f\,dA = \int_0^2 \int_{y^2}^4 f(x,y)\,dxdy$.

16. (a) The x limits are 0 and 2 (0 from the fact that $y = 0$ and $y = x^3$ intersect at (0,0); 2 was given). For each x in [0,2], y ranges from 0 to x^3. (Compare the vertical rectangle.) Thus $\iint_R f\,dA = \int_0^2 \int_0^{x^3} f(x,y)\,dydx$.

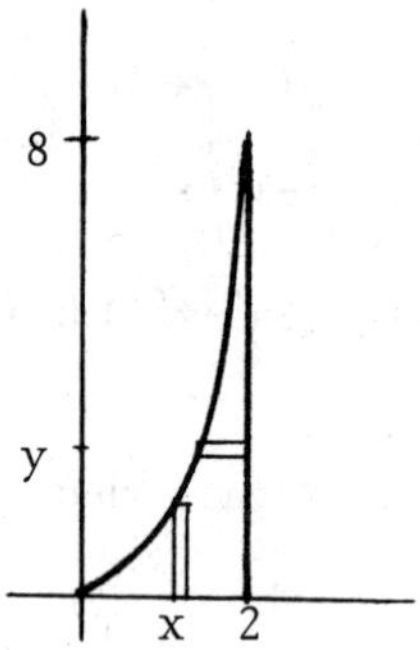

(b) The y limits are 0 and 8 (0 was given, 8 from the fact that the maximum value of the upper boundary function, x^3, is 8 for $0 \le x \le 2$). For each y in [0,8], x ranges from the curve, written as $x = \sqrt[3]{y}$, to $x = 2$.

Thus $\iint_R f\,dA = \int_0^8 \int_{\sqrt[3]{y}}^2 f(x,y)\,dx\,dy$.

NOTE: Analysis of a region as in #13 and #16 is very important to learn and master! It will be frequently encountered in multiple integration.

19. First, we must find the points of intersection of the curves. The lines intersect when $x+4 = 9-4x \iff 5x = 5 \iff x = 1$, i.e. at the point (1,5). The line $y = x+4$ and the curve $y = x^3/8$ intersect when $x+4 = x^3/8 \iff x^3-8x-32 = 0 \iff (x-4)(x^2+4x+8) = 0 \iff x = 4$, i.e. at (4,8). Finally, the line $y = 9-4x$ and the curve $y = x^3/8$ intersect when $9-4x = x^3/8 \iff x^3+32x-72 = 0 \iff (x-2)(x^2+2x+36) = 0 \iff x = 2$, i.e. at (2,1).

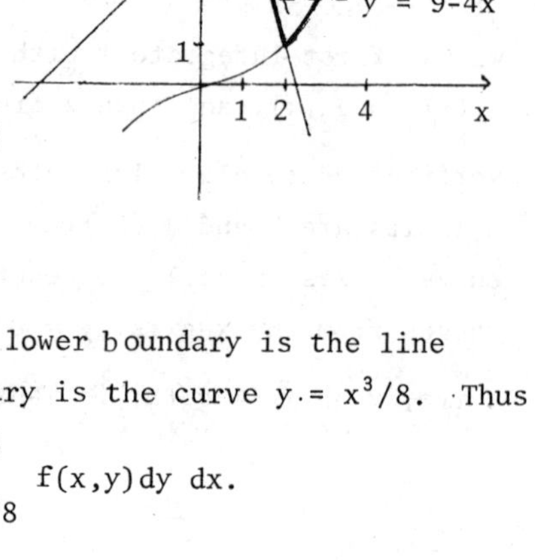

For the first part of the problem we note that the upper boundary of the region is the line $y = x+4$ always. If $1 \le x \le 2$, the lower boundary is the line $y = 9-4x$, and if $2 \le x \le 4$, the lower boundary is the curve $y = x^3/8$. Thus

$$\iint_R f(x,y)\,dA = \int_1^2 \int_{9-4x}^{x+4} f(x,y)\,dy\,dx + \int_2^4 \int_{x^3/8}^{x+4} f(x,y)\,dy\,dx.$$

For the second part of the problem we see that the right boundary is $y = x^3/8$, now written as $x = 2\sqrt[3]{y}$. If $1 \le y \le 5$, the left boundary is the line $y = 9-4x$ written in the form $x = (9-y)/4$. If $5 \le y \le 8$, the left boundary is the line $y = x+4$ now written as $x = y-4$. Thus

$$\iint_R f(x,y)\,dA = \int_1^5 \int_{(9-y)/4}^{2\sqrt[3]{y}} f(x,y)\,dx\,dy + \int_5^8 \int_{y-4}^{2\sqrt[3]{y}} f(x,y)\,dx\,dy.$$

22. Let ℓ_1, ℓ_2, and ℓ_3 denote the three lines given in the problem in that order. (Thus ℓ_1 is $x+2y = 5$, etc.). Then ℓ_1 and ℓ_2 intersect at $(3,1)$; ℓ_1 and ℓ_3 at $(-3,4)$; ℓ_2 and ℓ_3 at $(0,-2)$. For the first part of the problem, we see that if $-3 \le x \le 3$, the upper boundary is ℓ_1 written as $y = (5-x)/2$. If $-3 \le x \le 0$, the lower boundary is ℓ_3, $y = -2x-2$, and if $0 \le x \le 3$ the lower boundary is ℓ_2, $y = x-2$. Thus

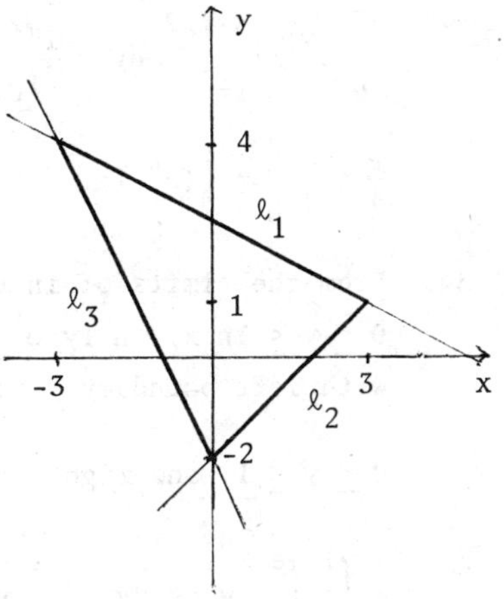

$$\iint_R f(x,y)\,dA = \int_{-3}^{0} \int_{-2x-2}^{(5-x)/2} f(x,y)\,dy\,dx + \int_0^3 \int_{x-2}^{(5-x)/2} f(x,y)\,dy\,dx.$$

For the second part of the problem, we see that if $-2 \le y \le 4$, the left boundary is ℓ_3, now written as $x = (-2-y)/2 = -1-y/2$. If $-2 \le y \le 1$, the right boundary is ℓ_2, $x = y+2$, and if $1 \le y \le 4$, the right boundary is ℓ_1, $x = 5-2y$.

Thus $$\iint_R f(x,y)\,dA = \int_{-2}^{1} \int_{-1-y/2}^{y+2} f(x,y)\,dx\,dy + \int_1^4 \int_{-1-y/2}^{5-2y} f(x,y)\,dx\,dy.$$

25. From the limits of integration, $R = \{(x,y)\colon -1 \le x \le 2,\ -\sqrt{4-x^2} \le y \le 4-x^2\}$. Thus the upper boundary is the parabola $y = 4-x^2$, and the lower boundary is the bottom half of the circle $x^2 + y^2 = 4$ for x between -1 and 2. The sketch is in the text.

28. From the limits: $R = \{(x,y)\colon -2 \le y \le -1,\ 3y \le x \le 2y\}$. Thus the right boundary is the line $x = 2y$ (or $y = x/2$), and the left boundary is the line $x = 3y$ (or $y = x/3$) for $-2 \le y \le 1$.

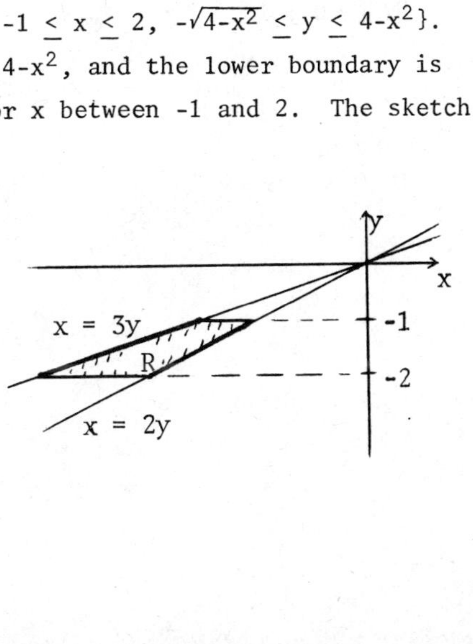

31. From the limits of integration $R = \{(x,y): 0 \le x \le 1, 2x \le y \le 2\}$, the triangular region shown--a Type I region. It is also a Type II region with left-hand boundary the y-axis, $x = 0$, and right-hand boundary the line $x = y/2$, $0 \le y \le 2$. Thus, by (17.14),

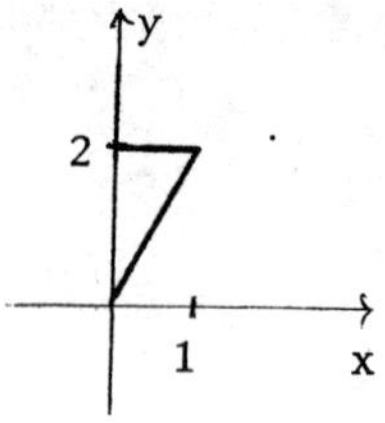

$$\int_0^1 \int_{2x}^2 e^{y^2}\, dy\, dx = \iint_R e^{-y^2}\, dA = \int_0^2 \int_0^{y/2} e^{y^2}\, dx\, dy =$$

$$\int_0^2 e^{y^2} x\Big]_{x=0}^{x=y/2} dy = \frac{1}{2}\int_0^2 e^{y^2} y\, dy = \frac{1}{4}\int_0^2 e^{y^2}(2y)\, dy =$$

$$\frac{1}{4} e^{y^2}\Big]_0^2 = \frac{1}{4}(e^4 - 1).$$

34. From the limits of integration, $R = \{(x,y): 1 \le x \le e, 0 \le y \le \ln x\}$, a Type I region. It is also of Type II with left boundary $y = \ln x$ now written as $x = e^y$, $0 \le y \le 1$, and right boundary $x = e$. Thus $\int_1^e \int_0^{\ln x} y\, dy dx$

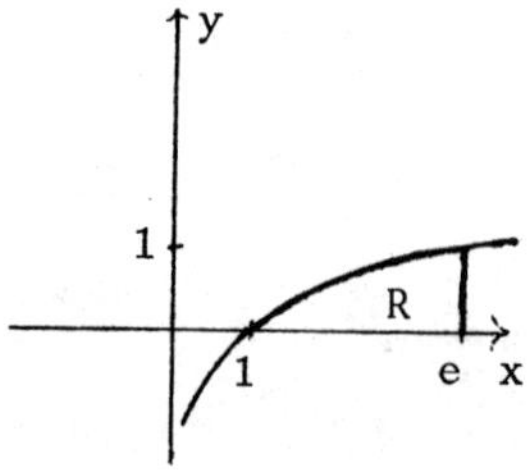

$$= \int_0^1 \int_{e^y}^e y\, dx\, dy = \int_0^1 y\, x\Big]_{e^y}^e dy = \int_0^1 y(e - e^y)\, dy = e\int_0^1 y\, dy$$

$$- \int_0^1 y\, e^y dy = \frac{ey^2}{2}\Big]_0^1 - (ye^y - e^y)\Big]_0^1 = (\frac{e}{2} - 0) - [(e-e)-(0-1)] = \frac{e}{2} - 1.$$

(Integration by parts, with $u = y$, $dv = e^y dy$, was used for $\int ye^y dy$.)

37. The order of integration in II of (17.14) is x 1st, then y. This would correspond to a summation 1st of $f(u_i, v_j)\Delta x_i \Delta y_j$ for rectangles in a horizontal strip, then summation over these horizontal strips. Thus

$$\lim_{\|\Delta\| \to 0} \sum_j \sum_i f(u_i v_j)\Delta x_i \Delta y_j .$$

EXERCISES 17.3, page 842

1. $A = \iint_R 1\, dA = \int_1^2 \int_{-x^2}^{1/x^2} 1\, dy\, dx =$

$\int_1^2 (\frac{1}{x^2} + x^2)dx = -\frac{1}{x} + \frac{x^3}{3}\Big]_1^2 =$

$\frac{13}{6} - (-\frac{2}{3}) = \frac{17}{6}$.

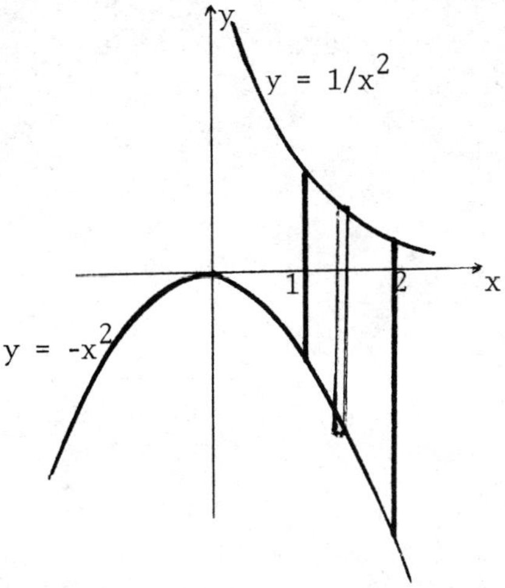

3. $A = \int_{-1}^{2} \int_{-y^2}^{y+4} dx\, dy.$

4. $A = \iint_R 1\, dA = \int_{-2}^{3} \int_{y-2}^{y^2} 1\, dx\, dy =$

$\int_{-2}^{3} (y^2-y+2)dy = \frac{y^3}{3} - \frac{y^2}{2} + 2y\Big]_{-2}^{3}$

$= \frac{21}{2} - (-\frac{26}{3}) = \frac{115}{6}$.

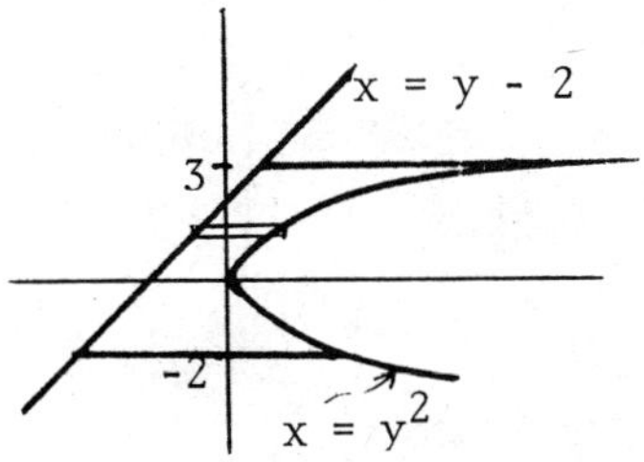

5. $A = \int_0^1 \int_x^{3x} dy\, dx + \int_1^2 \int_x^{4-x} dy\, dx.$

7. $A = \int_{-\pi}^{\pi} \int_{\sin x}^{e^x} 1\, dy\, dx = \int_{-\pi}^{\pi} (e^x - \sin x)dx$

$= e^x + \cos x\Big]_{-\pi}^{\pi} = (e^{\pi}-1)-(e^{-\pi}-1)$

$= e^{\pi} - e^{-\pi}.$

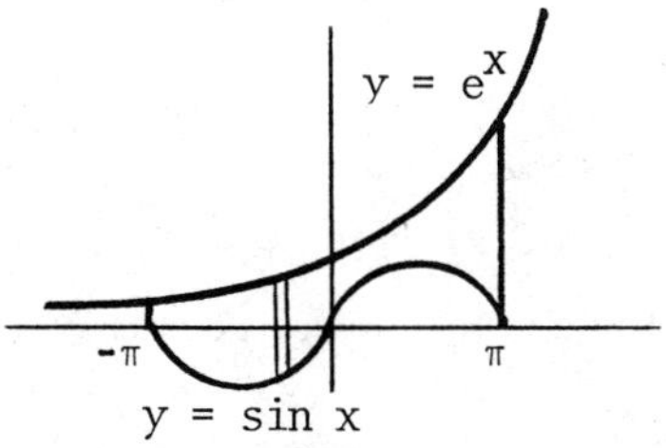

9. $A = \int_{-a}^{a} \int_{x^2}^{1/(1+x^2)} dy\, dx$ where $a = \sqrt{\frac{-1 + \sqrt{5}}{2}}$.

10. $A = \int_{-1}^{1} \int_{\sinh x}^{\cosh x} 1 \, dy \, dx =$

$\int_{-1}^{1} (\cosh x - \sinh x) dx =$

$\sinh x - \cosh x \Big]_{-1}^{1} = 2 \sinh 1.$

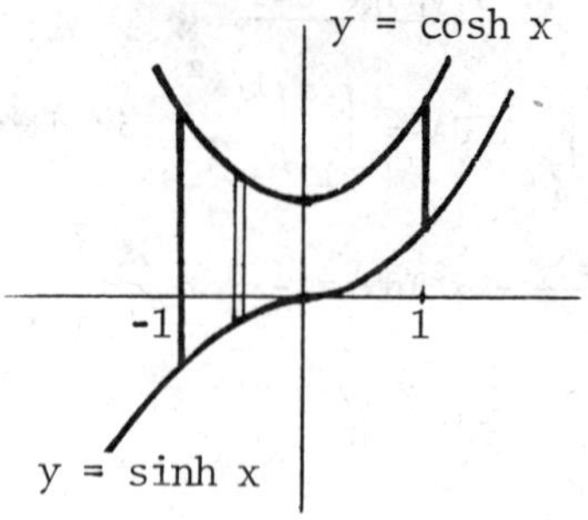

11. $V = \int_{0}^{2} \int_{0}^{1} (4x^2 + y^2) dy \, dx.$

13. The cylinder $x^2 + z^2 = 9$ has trace $x = 3$ in the 1st octant portion of the xy-plane. This cuts $y = 2x$ at $x = 3$, $y = 6$. Thus the solid is under the surface $z = \sqrt{9-x^2}$ and above the triangular region R as shown, $R = \{(x,y): 0 \le x \le 3, 0 \le y \le 2x\}$. Thus

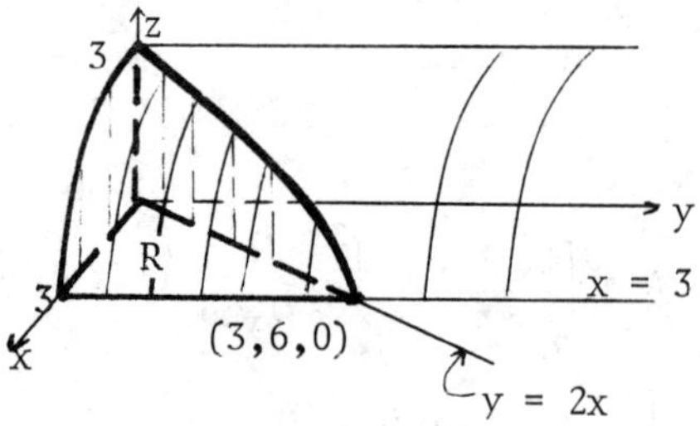

$$V = \iint_R \sqrt{9-x^2} \, dA = \int_{0}^{3} \int_{0}^{2x} (9-x^2)^{1/2} dy dx$$

$$= \int_{0}^{3} 2x(9-x^2)^{1/2} dx = -\frac{2}{3}(9-x^2)^{3/2}\Big]_{0}^{3} = \frac{2}{3} 9^{3/2} = 18.$$

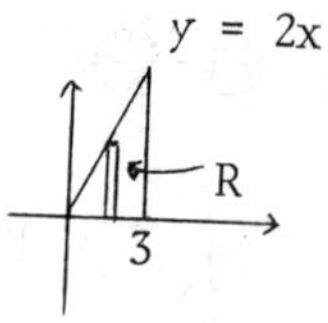

15. $V = \int_{0}^{2} \int_{0}^{4-2x} (4 - 2x - y) dy \, dx.$

16. The solid is below the surface $z = y^2$ and above the triangular region, R, in the 1st quadrant bounded by $y = x$ and $x = 4$: $R = \{(x,y): 0 \le x \le 4, 0 \le y \le x\}$. Thus $V = \int_{0}^{4} \int_{0}^{x} y^2 \, dy dx = \int_{0}^{4} \frac{y^3}{3}\Big]_{0}^{x} dx = \frac{1}{3} \int_{0}^{4} x^3 \, dx = \frac{64}{3}.$

17. $V = \int_{0}^{3} \int_{0}^{(6-2x)/3} (x^2+y^2) dy \, dx.$

19. The solid is below the surface $z = x^3$ and above the triangular region, R, in the 1st quadrant bounded by $y = x$ and $y = 1$: $R = \{(x,y): 0 \le x \le y, 0 \le y \le 1\}$. Thus $V = \int_{0}^{1} \int_{0}^{y} x^3 \, dx \, dy = \int_{0}^{1} \frac{x^4}{4}\Big]_{0}^{y} dy = \int_{0}^{1} \frac{y^4}{4} dy = \frac{1}{20}.$

21. $V = \int_{-1}^{2}\int_{2-x}^{4-x^2} (x^2+4)\,dy\,dx.$

22. The solid is below the plane $z - x - y = 4$, or $z = x + y + 4$, and above the region R in the xy plane bounded by the curves $y = x^3$ and $y = x^4$, which intersect at $x = 0$ and 1. Thus $V = \int_0^1\int_{x^4}^{x^3} (x+y+4)\,dy\,dx = \int_0^1 [xy + \frac{y^2}{2} + 4y]_{x^4}^{x^3}\,dx$

$= \int_0^1 (x^4 + \frac{x^6}{2} + 4x^3) - (x^5 + \frac{x^8}{2} + 4x^4)\,dx = \int_0^1 (-3x^4 + \frac{x^6}{2} + 4x^3 - x^5 - \frac{x^8}{2})\,dx$

$= -\frac{3}{5} + \frac{1}{14} + 1 - \frac{1}{6} - \frac{1}{18} = \frac{157}{630}.$

23. $V = 8\int_0^a\int_0^{\sqrt{a^2-x^2}} \sqrt{a^2-x^2}\,dy\,dx.$

28. The integrand, $\sqrt{x^2+y^2}$, indicates that the "roof" of the solid is the cone $z = \sqrt{x^2+y^2}$. (The "base" is, of course, the xy plane.) The region, R, in the xy-plane above which the solid extends is given by the limits of integration. $R = \{(x,y): 0 \le y \le 1, y^{1/2} \le x \le y^{1/3}\}$ and is bounded by $x = y^{1/2}$ (or $y = x^2$), $x = y^{1/3}$ (or $y = x^3$), $y = 0$ and $y = 1$.

EXERCISES 17.4, page 849

1. $M = \iint_R \rho\,dA = \int_0^9\int_0^{\sqrt{x}} (x+y)\,dydx = \int_0^9 xy + \frac{y^2}{2}\Big]_0^{\sqrt{x}}\,dx = \int_0^9 (x^{3/2} + \frac{x}{2})\,dx =$

$\frac{2}{5}x^{5/2} + \frac{x^2}{4}\Big]_0^9 = \frac{2}{5}(3^5) + \frac{9^2}{4} = 9^2(\frac{6}{5} + \frac{1}{4}) = \frac{81(29)}{20}.$

$M_y = \iint_R x\rho\,dA = \int_0^9\int_0^{\sqrt{x}} (x^2+xy)\,dy\,dx = \int_0^9 (x^{5/2} + \frac{x^2}{2})\,dx = \frac{2}{7}x^{7/2} + \frac{x^3}{6}\Big]_0^9 =$

$\frac{2}{7}\cdot 3^7 + \frac{9^3}{6} = \frac{729(43)}{42}$ and $\bar{x} = \frac{M_y}{M} = \frac{729(43)}{42}\cdot\frac{20}{81(29)} = \frac{1290}{203} \approx 6.4.$

$M_x = \iint_R y\,\rho\,dA = \int_0^9\int_0^{\sqrt{x}} (xy + y^2)\,dy\,dx = \int_0^9 (\frac{x^2}{2} + \frac{x^{3/2}}{3})\,dx = \frac{x^3}{6} + \frac{2}{15}x^{5/2}\Big]_0^9$

$= 9^2(\frac{9}{6} + \frac{6}{15}) = \frac{81(19)}{10}$ and $\bar{y} = \frac{M_x}{M} = \frac{19}{10}\cdot\frac{20}{29} = \frac{38}{29} \approx 1.3.$

2. See #16 below for sketch of region.

3. $M = \int_{-2}^{2}\int_{x^2}^{4} k|x|\,dy\,dx$, $M_y = k\int_{-2}^{2}\int_{x^2}^{4} x|x|\,dy\,dx$, $M_x = k\int_{-2}^{2}\int_{x^2}^{4} y|x|\,dy\,dx.$

4. To find the points of intersection we set $x^3 = 2x \Leftrightarrow x(x^2-2) = 0 \Rightarrow x = 0, \pm\sqrt{2}$. Since the distance of a point (x,y) from the x-axis is $|y|$, $\rho = k|y|$ where k is a constant. Thus for $-\sqrt{2} \le x \le 0$ since $y \le 0$ inside R, $\rho = -ky$, and for $0 \le x \le \sqrt{2}$, $\rho = ky$. Thus

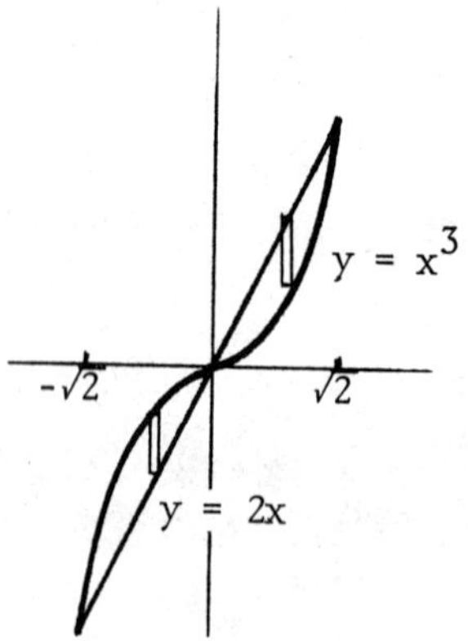

$$M = \iint_R \rho\, dA = \int_{-\sqrt{2}}^{0}\int_{2x}^{x^3} (-ky)\,dydx + \int_0^{\sqrt{2}}\int_{x^3}^{2x} (ky)\,dydx = \int_{-\sqrt{2}}^{0} \frac{k}{2}(4x^2-x^6)\,dx + \int_0^{\sqrt{2}} \frac{k}{2}(4x^2-x^6)\,dx \quad (= \int_{-\sqrt{2}}^{\sqrt{2}} (\ )\,dx\ )$$

$= \frac{k}{2}[\frac{4}{3}x^3 - \frac{x^7}{7}]_{-\sqrt{2}}^{\sqrt{2}} = \frac{32\sqrt{2}}{21}k$ (Note: from the symmetry of R about the origin and ρ about the x-axis, we could have just found the mass of the right half and doubled it.) $M_y = \iint_R x\rho\, dA$. The x does not alter this y integration as done above. The final result is only multiplied by x, and $M_y = \int_{-\sqrt{2}}^{\sqrt{2}} \frac{k}{2}(4x^3-x^7)\,dx = 0$ and $\overline{x} = 0$.

$M_x = \iint_R y\rho\, dA = \int_{-\sqrt{2}}^{0}\int_{2x}^{x^3} (-ky^2)\,dydx + \int_0^{\sqrt{2}}\int_{x^3}^{2x} (ky^2)\,dydx$, which, as above, reduces to $\frac{k}{3}\int_{-\sqrt{2}}^{\sqrt{2}} (8x^3-x^9)\,dx = 0$ and $\overline{y} = 0$ as well.

5. $M = \int_{-1}^{1}\int_0^{e^{-x^2}} |xy|\,dy\,dx = \int_{-1}^{1}\int_0^{e^{-x^2}} |x|y\,dy\,dx$ (since $y \ge 0$ in this region).

$M_y = \int_{-1}^{1}\int_0^{e^{-x^2}} x|x|y\,dy\,dx$, $M_x = \int_{-1}^{1}\int_0^{e^{-x^2}} |x|y^2\,dy\,dx$.

7. The region is the same as #4, Sec. 17.3. See the sketch there. Since $\rho = 1$, M is the same as the area, $M = 115/6$.

$$M_y = \iint_R x\rho\, dA = \int_{-2}^{3}\int_{y-2}^{y^2} x\,dxdy = \int_{-2}^{3} \frac{x^2}{2}\Big]_{y-2}^{y^2} dy = \frac{1}{2}\int_{-2}^{3} [y^4-(y-2)^2]\,dy =$$

$$\frac{1}{2}[\frac{y^5}{5} - \frac{(y-2)^3}{3}]_{-2}^{3} = \frac{1}{2}(\frac{724}{15} - \frac{224}{15}) = \frac{100}{6} \text{ and } \overline{x} = \frac{M_y}{M} = \frac{100}{6}\cdot\frac{6}{115} = \frac{20}{23}.$$

$$M_x = \iint_R y\rho\, dA = \int_{-2}^{3}\int_{y-2}^{y^2} y\,dx\,dy = \int_{-2}^{3} y(y^2-y+2)\,dy = \frac{y^4}{4} - \frac{y^3}{3} + y^2\Big]_{-2}^{3}$$

$$= \frac{81}{4} - \frac{32}{3} = \frac{115}{12} \text{ and } \bar{y} = \frac{M_x}{M} = \frac{115}{12} \cdot \frac{6}{115} = \frac{1}{2}.$$

9. $M = \int_{-\pi/4}^{\pi/4} \int_{1/2}^{\sec x} 4\,dy\,dx$, $M_y = \int_{-\pi/4}^{\pi/4} \int_{1/2}^{\sec x} 4x\,dy\,dx$, $M_x = \int_{-\pi/4}^{\pi/4} \int_{1/2}^{\sec x} 4y\,dy\,dx$.

10. Because of the shape of the region, it is better to integrate with respect to x first. Otherwise, the integrals would be needed for each of M, M_y and M_x. The upper x limit is obtained by solving $xy^2 = 1$ for x as a function of y. Thus $x = 1/y^2 = y^{-2}$.

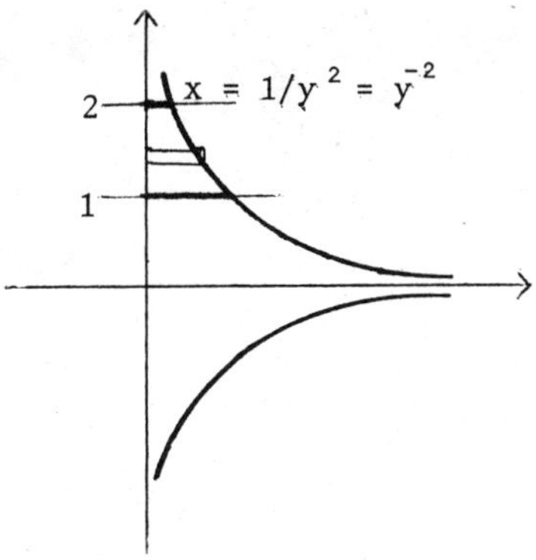

$$M = \iint_R \rho\,dA = \int_1^2 \int_0^{y^{-2}} (x^2+y^2)\,dx\,dy =$$

$$\int_1^2 \left[\frac{x^3}{3} + y^2x\right]_0^{y^{-2}} dy = \int_1^2 \left(\frac{y^{-6}}{3} + y^2y^{-2}\right)dy =$$

$$\int_1^2 \left(\frac{y^{-6}}{3} + 1\right)dy = -\frac{y^{-5}}{15} + y\Big]_1^2 = \left(-\frac{1}{15(32)} + 2\right) -$$

$$\left(-\frac{1}{15} + 1\right) = 1 + \frac{1}{15}\left(1 - \frac{1}{32}\right) = 1 + \frac{31}{480} = \frac{511}{480}.$$

$$M_y = \iint_R x\rho\,dA = \int_1^2 \int_0^{y^{-2}} x(x^2+y^2)\,dx\,dy = \int_1^2 \int_0^{y^{-2}} (x^3+xy^2)\,dx\,dy =$$

$$\int_1^2 \left[\frac{x^4}{4} + \frac{x^2y^2}{2}\right]_0^{y^{-2}} dy = \int_1^2 \left(\frac{y^{-8}}{4} + \frac{y^{-4}y^2}{2}\right)dy = \int_1^2 \left(\frac{y^{-8}}{4} + \frac{y^{-2}}{2}\right)dy = -\frac{y^{-7}}{28} - \frac{y^{-1}}{2}\Big]_1^2$$

$$= -\frac{1}{28y^7} - \frac{1}{2y}\Big]_1^2 = \left(-\frac{1}{28(128)} - \frac{1}{4}\right) - \left(-\frac{1}{28} - \frac{1}{2}\right) = \frac{1}{4} + \frac{1}{28}\left(1 - \frac{1}{128}\right) = \frac{1}{4}\left(1 + \frac{127}{896}\right)$$

$$= \frac{1023}{3584}. \text{ Thus } \bar{x} = \frac{M_y}{M} = \frac{1023}{3584} \cdot \frac{480}{511} = \frac{(1023)(15)}{(112)(511)} = \frac{15{,}345}{57{,}232} \approx 0.268.$$

$$M_x = \iint_R y\rho\,dA = \int_1^2 \int_0^{y^{-2}} (x^2y + y^3)\,dx\,dy = \int_1^2 \left(\frac{yy^{-6}}{3} + y^3y^{-2}\right)dy = \int_1^2 \left(\frac{y^{-5}}{3} + y\right)dy$$

$$= \frac{y^{-4}}{12} + \frac{y^2}{2}\Big]_1^2 = \left(-\frac{1}{12(16)} + 2\right) - \left(-\frac{1}{12} + \frac{1}{2}\right) = \frac{3}{2} + \frac{1}{12}\left(1 - \frac{1}{16}\right) = \frac{3}{2}\left(1 + \frac{5}{96}\right) = \frac{303}{192}$$

$$= \frac{101}{64}. \text{ Thus } \bar{y} = \frac{M_x}{M} = \frac{101}{64} \cdot \frac{480}{511} = \frac{(101)(15)}{(2)(511)} = \frac{1515}{1022} \approx 1.482.$$

11. $M = \int_0^1 \int_0^x x\,dy\,dx + \int_1^2 \int_0^{1/x} x\,dy\,dx$, $M_y = \int_0^1 \int_0^x x^2dy\,dx + \int_1^2 \int_0^{1/x} x^2dy\,dx$,

$$M_x = \int_0^1 \int_0^x yx\,dy\,dx + \int_1^2 \int_0^{1/x} yx\,dy\,dx.$$

13. $M = \iint_R \rho\, dA = \int_0^1 \int_0^{e^{-x}} y^2 dy\, dx = \int_0^1 \frac{y^3}{3}\Big]_0^{e^{-x}} dx = \frac{1}{3}\int_0^1 e^{-3x}\, dx = -\frac{1}{9} e^{-3x}\Big]_0^1$

$= \frac{1-e^{-3}}{9}$. $M_y = \iint_R x\, \rho\, dA = \int_0^1 \int_0^{e^{-x}} xy^2 dy\, dx = \frac{1}{3}\int_0^1 xe^{-3x}\, dx$. Integrating by parts, with $u = x$, $dv = e^{-3x}dx$, we obtain $\int xe^{-3x}dx = -\frac{1}{3}xe^{-3x} - \frac{1}{9}e^{-3x} + C$

$= -\frac{1}{9}e^{-3x}(3x+1) + C$. Thus, $M_y = \frac{1}{3}[-\frac{1}{9}e^{-3x}(3x+1)]_0^1 = \frac{1}{3}[-\frac{4}{9}e^{-3} + \frac{1}{9}]$

$= \frac{1-4e^{-3}}{27}$, and $\bar{x} = \frac{M_y}{M} = \frac{1-4e^{-3}}{27} \cdot \frac{9}{1-e^{-3}}$.

$M_x = \iint_R y\, \rho\, dA = \int_0^1 \int_0^{e^{-x}} y^3 dy\, dx = \int_0^1 \frac{y^4}{4}\Big]_0^{e^{-x}} dx = \frac{1}{4}\int_0^1 e^{-4x}dx = -\frac{1}{16}e^{-4x}\Big]_0^1$

$= \frac{1-e^{-4}}{16}$. Thus $\bar{y} = \frac{M_x}{M} = \frac{1-e^{-4}}{16} \cdot \frac{9}{1-e^{-3}}$.

15. The limits of integration and the density are the same as in Exercise 1. The integrand for I_x is $xy^2 + y^3$, and that of I_y is $x^3 + x^2y$. Then for I_o, use $I_o = I_x + I_y$.

16. $I_x = \iint_R y^2\, \rho\, dA = \int_0^8 \int_0^{x^{1/3}} y^2(y^2)\, dydx =$

$\frac{1}{5}\int_0^8 x^{5/3}\, dx = \frac{1}{5}(\frac{3}{8})x^{8/3}\Big]_0^8 = \frac{3}{40} \cdot 2^8 = \frac{96}{5}$.

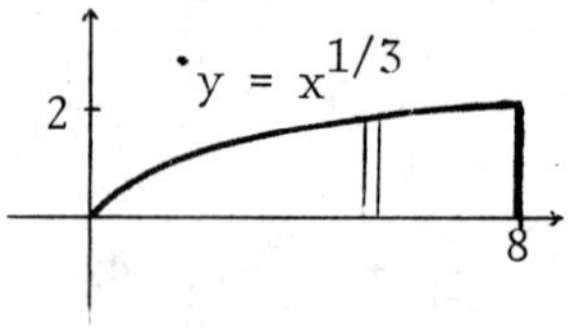

$I_y = \iint_R x^2\, \rho\, dA = \int_0^8 \int_0^{x^{1/3}} x^2y^2 dydx = \frac{1}{3}\int_0^8 x^3 dx = \frac{8^4}{12} = \frac{1024}{3}$.

$I_o = I_x + I_y = \frac{96}{5} + \frac{1024}{3} = \frac{5408}{15}$.

17. The integrands are $ky^2|x|$ for I_x and $kx^2|x|$ for I_y. See #3 above for the limits of integration.

19. (a) Here, for convenience, we place the origin at one corner and the axes coincident with 2 sides. Then either I_x or I_y will serve. Since the lamina is homogeneous, $\rho(x,y) = \rho$, a constant, and

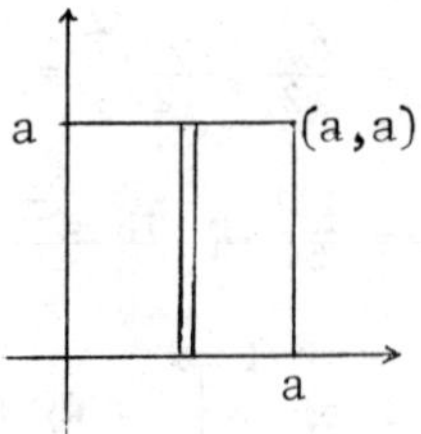

$I_x = \iint_R y^2\, \rho\, dA = \rho \int_0^a \int_0^a y^2\, dydx = \rho \cdot \frac{a^3}{3}\int_0^a dx = \frac{\rho}{3} a^4$.

(b) In this part it is more convenient to place the origin at the center and the axes along the diagonals of the square. The equations of the 2 upper edges are $y = \frac{a}{\sqrt{2}} \pm x$, and those of the 2 lower edges are $y = -\frac{a}{\sqrt{2}} \pm x$. Because of the symmetry, either diagonal may be chosen, i.e. $I_x = I_y$. Because of the change in upper and lower boundaries at x = 0, the integral over R must be split into the sum of integrals over the left and right halves. Thus

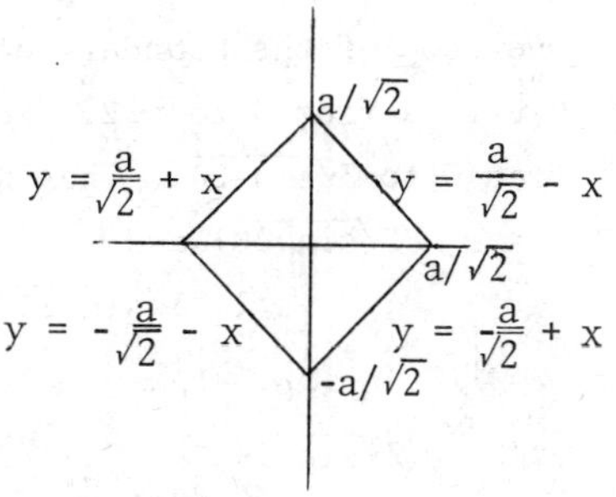

$$I_y = \iint_R x^2 \rho \, dA = \rho \int_{-\frac{a}{\sqrt{2}}}^{0} \int_{\frac{-a}{\sqrt{2}} - x}^{\frac{a}{\sqrt{2}} + x} x^2 \, dy \, dx + \rho \int_{0}^{a/\sqrt{2}} \int_{\frac{-a}{\sqrt{2}} + x}^{\frac{a}{\sqrt{2}} - x} x^2 \, dy \, dx =$$

$$\rho \int_{-a/\sqrt{2}}^{0} x^2 \left(\frac{2a}{\sqrt{2}} + 2x\right) dx + \rho \int_{0}^{a/\sqrt{2}} x^2 \left(\frac{2a}{\sqrt{2}} - 2x\right) dx = 2\rho\left[\frac{ax^3}{3\sqrt{2}} + \frac{x^4}{4}\right]_{-a/\sqrt{2}}^{0} +$$

$$2\rho\left[\frac{ax^3}{3\sqrt{2}} - \frac{x^4}{4}\right]_{0}^{a/\sqrt{2}} = 2\rho\left[- \frac{-a^4}{3\cdot\sqrt{2}^4} - \frac{a^4}{4\sqrt{2}^4}\right] + 2\rho\left[\frac{a^4}{3\sqrt{2}^4} - \frac{a^4}{4\sqrt{2}^4}\right] =$$

$$\frac{2\rho a^4}{4}\left(\frac{1}{3} - \frac{1}{4} + \frac{1}{3} - \frac{1}{4}\right) = \frac{\rho a^4}{12}.$$

(c) Since $I_x = I_y = \frac{\rho a^4}{12}$, $I_o = I_x + I_y = \frac{\rho a^4}{6}$.

21. (#21, relating to #19 above, will be included rather than #22.) Using I_x from #19(a), we seek $\bar{\bar{y}}$ such that $M\bar{\bar{y}}^2 = I_x$ where M = mass = ρ•Area $= \rho a^2$. Thus $\bar{\bar{y}}^2 = \frac{I_x}{M} = \frac{\rho}{3} a^4 \cdot \frac{1}{\rho a^2} = \frac{a^2}{3}$, and $\bar{\bar{y}} = \frac{a}{\sqrt{3}}$.

23. $I_x = \int_{-a}^{a} \int_{-\sqrt{a^2-x^2}}^{\sqrt{a^2-x^2}} \rho y^2 \, dy \, dx$, where the circle has been centered at the origin, and the diameter along the x-axis has been selected.

EXERCISES 17.5, page 856

1. One loop of the boundary of R is traced out as θ varies from 0 to $\pi/2$. For such θ, r ranges from 0 to $3\sqrt{\sin 2\theta}$. Thus $R = \{(r,\theta): 0 \leq \theta \leq \pi/2, 0 \leq r \leq 3\sqrt{\sin 2\theta}\}$ and

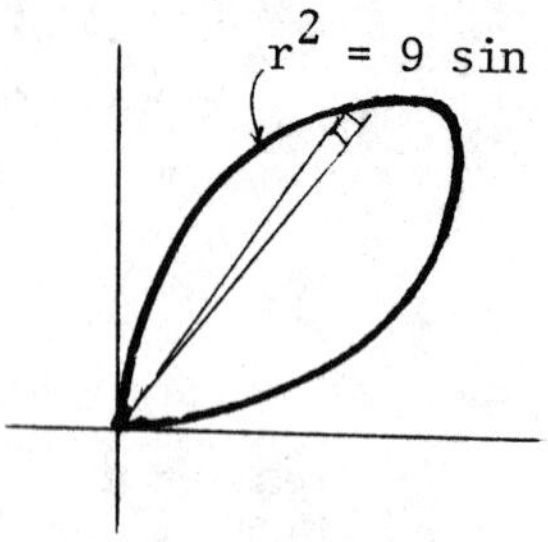

$$A = \iint_R dA = \int_0^{\pi/2}\int_0^{3\sqrt{\sin 2\theta}} r\,dr\,d\theta =$$

$$\int_0^{\pi/2} \frac{r^2}{2}\Big]_0^{3\sqrt{\sin 2\theta}} d\theta = \frac{9}{2}\int_0^{\pi/2} \sin 2\theta\,d\theta =$$

$$\frac{9}{4}[-\cos 2\theta]_0^{\pi/2} = \frac{9}{2}.$$

3. $A = \int_{2\pi/3}^{4\pi/3}\int_3^{2-2\cos\theta} r\,dr\,d\theta = 2\int_{2\pi/3}^{\pi}\int_3^{2-2\cos\theta} r\,dr\,d\theta.$

4. Since $R = \{(r,\theta): 0 \leq \theta \leq 2\pi, 0 \leq r \leq 3 + 2\sin\theta\}$,

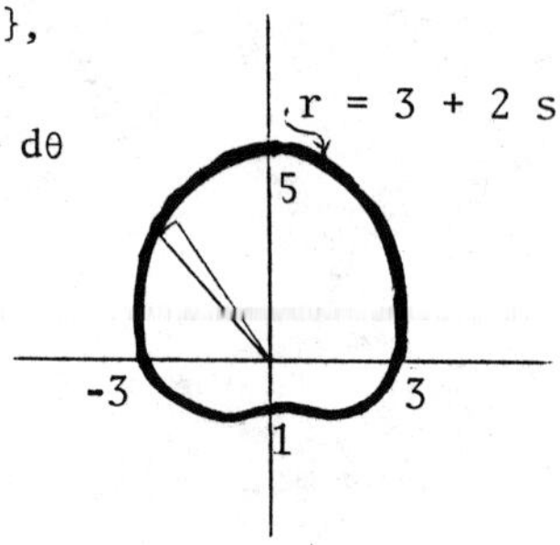

$$A = \int_0^{2\pi}\int_0^{3+2\sin\theta} r\,dr\,d\theta = \int_0^{2\pi} \frac{(3 + 2\sin\theta)^2}{2} d\theta$$

$$= \int_0^{2\pi} \frac{9 + 12\sin\theta + 4\sin^2\theta}{2} d\theta =$$

$$= \frac{1}{2}\int_0^{2\pi} (11 + 12\sin\theta - 2\cos 2\theta)d\theta$$

$$= \frac{1}{2}[11\theta - 12\cos\theta - \sin 2\theta]_0^{2\pi} = 11\pi.$$

5. With $\rho(r,\theta) = kr$, $M = \int_{-\pi/2}^{\pi/2}\int_0^{2\cos\theta} (kr)r\,dr\,d\theta$, $M_y = \iint_R x\,\rho\,dA =$

$$\int_{-\pi/2}^{\pi/2}\int_0^{2\cos\theta} (r\cos\theta)(kr)r\,dr\,d\theta,\ M_x = \iint_R y\,\rho\,dA =$$

$$\int_{-\pi/2}^{\pi/2}\int_0^{2\cos\theta} (r\sin\theta)(kr)r\,dr\,d\theta.$$

7. The values of θ at the points of intersection satisfy $3 \sin \theta = 1 + \sin \theta$ or $\sin \theta = 1/2$. Thus $\theta = \pi/6$ and $5\pi/6$. Thus $R = \{(r,\theta): \pi/6 \leq \theta \leq 5\pi/6,\ 1 + \sin \theta \leq r \leq 3 \sin \theta\}$. The density is given by $\rho(r,\theta) = k/r$ where k is a constant. By symmetry of R and independence of ρ on θ, the total mass is twice the mass of the right half where $\pi/6 \leq \theta \leq \pi/2$. Thus

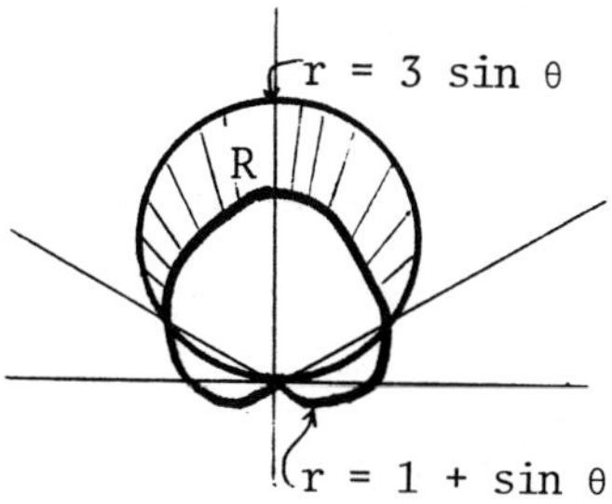

$$M = 2\int_{\pi/6}^{\pi/2}\int_{1+\sin\theta}^{3\sin\theta} \frac{k}{r}\cdot r\,dr\,d\theta = 2k\int_{\pi/6}^{\pi/2} [3 \sin \theta - (1 + \sin \theta)]d\theta =$$

$$2k\int_{\pi/6}^{\pi/2} (2 \sin \theta - 1)d\theta = 2k[-2 \cos \theta - \theta]_{\pi/6}^{\pi/2} = 2k[(0 - \tfrac{\pi}{2})+(2 \cos \tfrac{\pi}{6} + \tfrac{\pi}{6})] =$$

$2k(\sqrt{3} - \frac{\pi}{3})$. For the center of mass, we 1st observe that by symmetry (or by direct calculation of M_y), $\bar{x} = 0$. Also,

$$M_x = \iint_R y\rho\, dA = \int_{\pi/6}^{5\pi/6}\int_{1+\sin\theta}^{3\sin\theta} (r \sin \theta)(\tfrac{k}{r})\cdot r\, dr\, d\theta$$

$$= k\int_{\pi/6}^{5\pi/6} (\sin \theta) \left.\frac{r^2}{2}\right]_{1+\sin\theta}^{3\sin\theta} d\theta = \frac{k}{2}\int_{\pi/6}^{5\pi/6} (8 \sin^3\theta - \sin \theta - 2 \sin^2\theta)d\theta,$$

and using $\sin^2\theta = 1 - \cos^2\theta = \dfrac{1 - \cos 2\theta}{2}$,

$$M_x = \frac{k}{2}\int_{\pi/6}^{5\pi/6} (8(1 - \cos^2\theta)\sin \theta - \sin \theta - 1 + \cos 2\theta)d\theta$$

$$= \frac{k}{2}[-7 \cos \theta + \frac{8}{3} \cos^3\theta - \theta + \frac{1}{2} \sin 2\theta]_{\pi/6}^{5\pi/6}$$

$$= \frac{k}{2}[(\frac{7\sqrt{3}}{2} + \frac{8}{3}(-\frac{\sqrt{3}}{2})^3 - \frac{5\pi}{6} + \frac{1}{2}(-\frac{\sqrt{3}}{2}))-(-\frac{7\sqrt{3}}{2} + \frac{8}{3}(\frac{\sqrt{3}}{2})^3 - \frac{\pi}{6} + \frac{1}{2}(\frac{\sqrt{3}}{2})]$$

$$= \frac{k}{2}(5\sqrt{3} - \frac{2\pi}{3} - \frac{\sqrt{3}}{2}) = \frac{k}{2}[\frac{9\sqrt{3}}{2} - \frac{2\pi}{3}]$$

and $\bar{y} = \dfrac{M_x}{M} = k(\frac{9\sqrt{3}}{4} - \frac{\pi}{3})/2k(\sqrt{3} - \frac{\pi}{3}) \approx \dfrac{2.85}{1.37} \approx 2.08$.

9. With $\rho(r,\theta) = \rho$, $I_0 = \iint_R (x^2+y^2)\rho\, dA = \iint_R r^2 \rho\, dA = \rho\int_0^{\pi/2}\int_0^{\sqrt{a \sin 2\theta}} (r^2)r\, dr\, d\theta$

$$+ \rho\int_{\pi}^{3\pi/2}\int_0^{\sqrt{a \sin 2\theta}} (r^2)r\, dr\, d\theta = 2\rho\int_0^{\pi/2}\int_0^{\sqrt{a \sin 2\theta}} r^3\, dr\, d\theta.$$

10. The region $R = \{(r,\theta): \frac{1}{r} \le \theta \le \frac{2}{r},$ $1 \le r \le 2\}$ is shown. Let the constant density be k. Then with $x^2 + y^2 = r^2$ we obtain

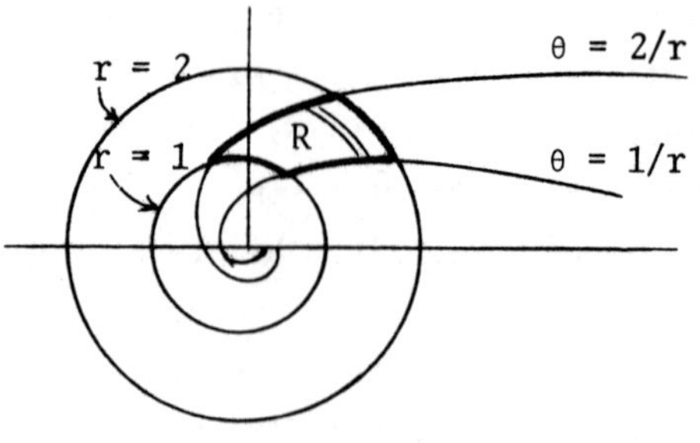

$$I_o = \iint_R (x^2+y^2)\rho\, dA =$$

$$\int_1^2 \int_{1/r}^{2/r} r^2 \cdot k \cdot r\, d\theta\, dr =$$

$$k\int_1^2 r^3\left(\frac{2}{r} - \frac{1}{r}\right) dr = k\int_1^2 r^2 dr = k\cdot\frac{7}{3}$$

11. With $\rho(r,\theta) = \rho$ and the circle centered at the origin with the x-axis as the selected line, $I_x = \iint_R y^2\ \rho\ dA = \int_0^{2\pi}\int_0^a (r \sin\theta)^2 \cdot \rho \cdot r\, dr\, d\theta.$

13. The region is bounded by the upper half of the circle $x^2 + y^2 = a^2$ and the x-axis since $0 \le y \le \sqrt{a^2-x^2} \Longrightarrow (x,y)$ is on or within this circle and above the x-axis. Thus in polar coordinates $R = \{(r,\theta): 0 \le r \le a,\ 0 \le \theta \le \pi\}$ and

$$\int_{-a}^{a}\int_0^{\sqrt{a^2-x^2}} e^{-(x^2+y^2)}\, dydx = \int_0^{\pi}\int_0^a e^{-r^2}\ r\ dr\ d\theta = \int_0^{\pi}\left[-\frac{1}{2}e^{-r^2}\right]_0^a d\theta =$$

$\frac{\pi}{2}(1 - e^{-a^2})$. (Note: This integral is impossible to evaluate exactly in rectangular coordinates, but it is trivial in polar.)

16. R is bounded by the upper half of the circle $x^2+y^2 = 1$ (since $0 \le y \le \sqrt{1-x^2}$) but only in the 1st quadrant portion (since $0 \le x \le 1$). Thus

$$\int_0^1\int_0^{\sqrt{1-x^2}} e^{\sqrt{x^2+y^2}}\, dydx = \int_0^{\pi/2}\int_0^1 e^r\ r\ dr\ d\theta.$$ Let $u = r$, $dv = e^r dr$ so that

$du = dr$, $v = e^r$. Then $\int_0^1 e^r r\ dr = re^r\Big]_0^1 - \int_0^1 e^r\ dr = e - e^r\Big]_0^1 = e-(e-1) = 1,$

and the integral reduces to $\int_0^{\pi/2} 1\ d\theta = \frac{\pi}{2}$.

17. $V = 8\int_0^{\pi/2}\int_3^5 \sqrt{25-r^2}\ r\ dr\ d\theta$. (The integral is the volume of the 1st octant portion of the solid. By symmetry, this can be multiplied by 8 to obtain the entire volume. Hence the 8 in front of the integral.)

19. The upper half of the solid lies above the region R inside the circle $r = 2\cos\theta$, $R = \{(r,\theta): -\pi/2 \le \theta \le \pi/2,\ 0 \le r \le 2\cos\theta\}$, and the top of the solid is the cone $z = f(r,\theta) = r$. Thus, for the upper half,

$$V = \int_{-\pi/2}^{\pi/2}\int_0^{2\cos\theta} (r)\, r\, dr\, d\theta = \frac{8}{3}\int_{-\pi/2}^{\pi/2} \cos^3\theta\, d\theta = \frac{8}{3}\int_{-\pi/2}^{\pi/2} (1 - \sin^2\theta)\cos\theta\, d\theta$$

$$= \frac{8}{3}\left[\sin\theta - \frac{\sin^3\theta}{3}\right]_{-\pi/2}^{\pi/2} = \frac{8}{3}\cdot\frac{4}{3} = \frac{32}{9}.$$ The lower half has the same volume (by symmetry), and the total volume is, therefore, $2(\frac{32}{9}) = \frac{64}{9}$.

EXERCISES 17.6, page 864

1. In Example 1, the order of integration was z, x, y. If we use the order z, y, x, then the 1st z integration is the same as there and, if I denotes the integral,

$$I = \int_{-1}^{3}\int_1^4 8xy^3\, dydx = \int_{-1}^{3} 2xy^4\Big]_1^4 dx = \int_{-1}^{3} 2(255)x\, dx = 255\, x^2\Big]_{-1}^{3} = 255(8).$$

If we integrate in the order x, y, z or x, y, the 1st integration is the same in both cases:

$$\int_{-1}^{3} 3xy^3z^2dx = \frac{3}{2}\, y^3z^2x^2\Big]_{-1}^{3} = 12y^3z^3,$$ and $$\int_0^2\int_1^4\int_{-1}^{3} 3xy^3z^2\, dxdydz =$$

$$\int_0^2\int_1^4 12y^3z^2\, dydz = \int_0^2 3(255)z^2dz = 255(8),\quad \int_0^2\int_1^4\int_{-1}^{3} 3xy^3z^2\, dxdzdy =$$

$$\int_1^4\int_0^2 12y^3z^2\, dzdy = \int_1^4 4(8)y^3dy = 8(255).$$

If the order is y, x, z or y, z, x, the 1st integration is the same for both:

$$\int_1^4 3xy^3z^2\, dy = \frac{3}{4}xz^2y^4\Big]_1^4 = \frac{3}{4}(255)xz^2$$ and again both integrals work out to 255(8).

4. $$\int_1^2\int_0^{z^2}\int_{x+z}^{x-z} z\, dydxdz = \int_1^2\int_0^{z^2} zy\Big]_{x+z}^{x-z} dx\, dz = \int_1^2\int_0^{z^2} z[(x-z)-(x+z)]\, dxdz =$$

$$\int_1^2\int_0^{z^2} (-2z^2)\, dxdz = \int_1^2 (-2z^2)x\Big]_0^{z^2} dz = \int_1^2 (-2z^4)dz = -\frac{2}{5}(2^5-1) = \frac{-62}{5}.$$

NOTE: In solving problems involving different orders of integration in triple integrals, it may be helpful to follow these steps: 1. Sketch the region Q. 2. For the variable to be integrated 1st, determine how it varies within Q (e.g. z will range from the lower to the upper surfaces, expressed as functions of x and y; y would vary from the left face to the right, expressed in terms of x and z; x would vary from the back face to the front, expressed in terms of y and z). Sketching a column parallel to the axis of this 1st variable may help. 3. The remaining 2 variables are now seen to vary in a region R in a coordinate plane. R will be in the xy-plane if z was integrated 1st, etc. (R can be thought of as the "projection" of Q onto that plane.) The limits in R are then found as in the case of double integrals, Sec. 17.2.

7. (a) <u>z integrated first.</u> In Q, z varies from the xy-plane, z = 0, to the slant plane written as $z = \frac{1}{3}(6-x-2y)$. The region R in the xy-plane is bounded by the axes and the line $x+2y = 6$ (the trace of the plane $x+2y+3z = 6$ in the xy-plane). Thus $R = \{(x,y): 0 \le x \le 6,\ 0 \le y \le 1/2(6-x)\}$ and $R = \{(x,y): 0 \le y \le 3,\ 0 \le x \le 6-2y\}$, and we obtain:

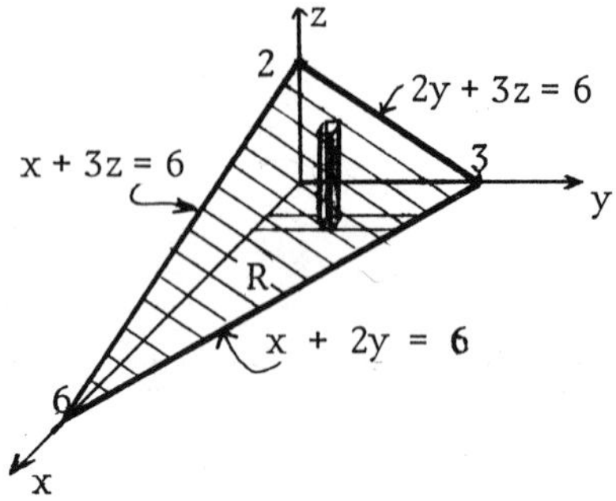

$$\int_0^6 \int_0^{\frac{1}{2}(6-x)} \int_0^{\frac{1}{3}(6-x-2y)} f(x,y,z)\ dzdydx$$

and $\displaystyle\int_0^3 \int_0^{6-2y} \int_0^{\frac{1}{3}(6-x-2y)} f(x,y,z)\ dzdxdy.$

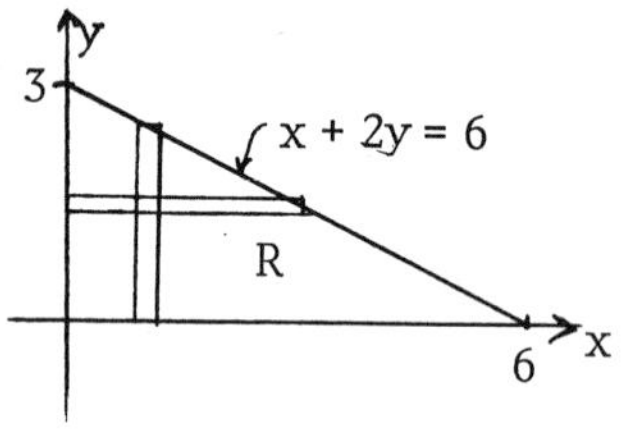

(b) <u>x integrated first.</u> In Q, x varies from the yz-plane, x = 0, to the slant plane now written as $x = 6-2y-3z$. R is in the yz-plane bounded by the axes and the line $2y+3z = 6$, and $R = \{(y,z): 0 \le y \le 3,\ 0 \le z \le 1/3(6-2y)\}$ and $R = \{(y,z): 0 \le z \le 2,\ 0 \le y \le 1/2(6-3z)\}$

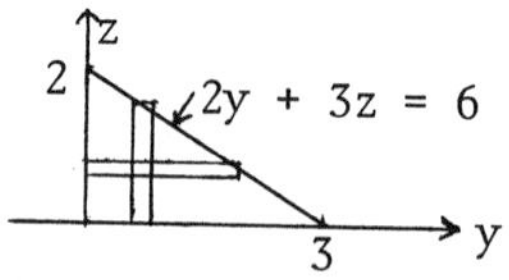

and we obtain $\displaystyle\int_0^3 \int_0^{\frac{1}{3}(6-2y)} \int_0^{6-2y-3z} f(x,y,z)\ dxdzdy$ and

$$\int_0^2 \int_0^{\frac{1}{2}(6-3z)} \int_0^{6-2y-3z} f(x,y,z)\ dxdydz.$$

(c) <u>y integrated first.</u> In Q, y varies from y = 0, to the slant plane $y = \frac{1}{2}(6-3z-x)$. $R = \{(x,z): 0 \le x \le 6,\ 0 \le z \le \frac{1}{3}(6-x)\}$ in the xz-plane from which the answers follow as above.

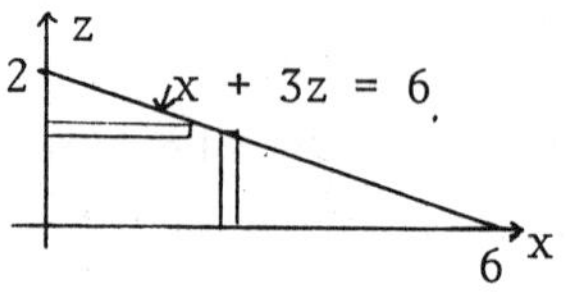

10. (a) z first. In the ellipsoid Q, z varies from the bottom surface, $z = -\frac{1}{2}\sqrt{36-36x^2-9y^2}$ to the top, $z = \frac{1}{2}\sqrt{36-36x^2-9y^2}$. R is the region in the xy-plane bounded by the ellipse $36x^2+9y^2 = 36$ or $x^2 + \frac{y^2}{4} = 1$. Thus R = $\{(x,y)\colon -1 \le x \le 1,\ -2\sqrt{1-x^2} \le y \le 2\sqrt{1-x^2}\}$ = $\{(x,y)\colon -2 \le y \le 2,\ -\sqrt{1-\frac{y^2}{4}} \le x \le \sqrt{1-\frac{y^2}{4}}\}$.

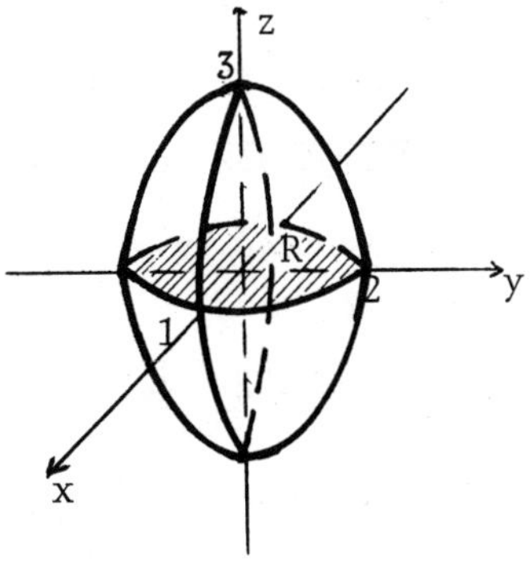

Thus we get $\displaystyle\int_{-1}^{1}\int_{-2\sqrt{1-x^2}}^{2\sqrt{1-x^2}}\int_{-\frac{1}{2}\sqrt{36-36x^2-9y^2}}^{\frac{1}{2}\sqrt{36-36x^2-9y^2}} f(x,y,z)\,dzdydx,$

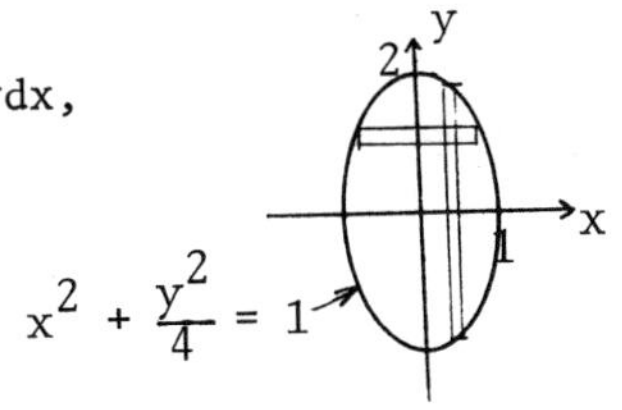

$$\int_{-2}^{2}\int_{-\sqrt{1-\frac{y^2}{4}}}^{\sqrt{1-\frac{y^2}{4}}}\int_{-\frac{1}{2}\sqrt{36-36x^2-9y^2}}^{\frac{1}{2}\sqrt{36-36x^2-9y^2}} f(x,y,z)\ dzdxdy.$$

(b) y first. In Q, y varies between the left and right surfaces $y = \mp\frac{1}{3}\sqrt{36-36x^2-4z^2}$. R is the region in the xz-plane bounded by the ellipse $9x^2+z^2 = 9$. Thus R = $\{(x,z)\colon -1 \le x \le 1,$ $-3\sqrt{1-x^2} \le z \le 3\sqrt{1-x^2}\}$ = $\{(x,z)\colon -3 \le z \le 3,$ $-\frac{1}{3}\sqrt{9-z^2} \le x \le \frac{1}{3}\sqrt{9-z^2}$ and we get

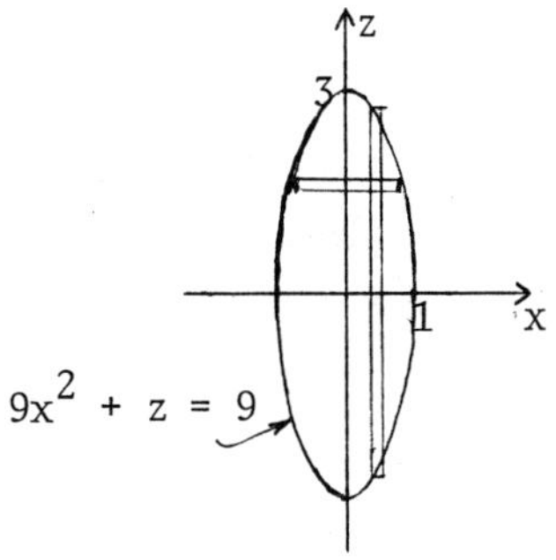

$$\int_{-1}^{1}\int_{-3\sqrt{1-x^2}}^{3\sqrt{1-x^2}}\int_{-\frac{1}{3}\sqrt{36-36x^2-4z^2}}^{\frac{1}{3}\sqrt{36-36x^2-4z^2}} f(x,y,z)\ dydzdx \text{ and}$$

$$\int_{-3}^{3}\int_{-\frac{1}{3}\sqrt{9-z^2}}^{\frac{1}{3}\sqrt{9-z^2}}\int_{-\frac{1}{3}\sqrt{36-36x^2-4z^2}}^{\frac{1}{3}\sqrt{36-36x^2-4z^2}} f(x,y,z)\ dydxdz.$$

(c) x first. In Q, x varies from the back to the front surfaces $x = \pm \frac{1}{6}\sqrt{36-9y^2-4z^2}$ and R is in the yz-plane bounded by $9y^2+4z^2 = 36$. As above we get

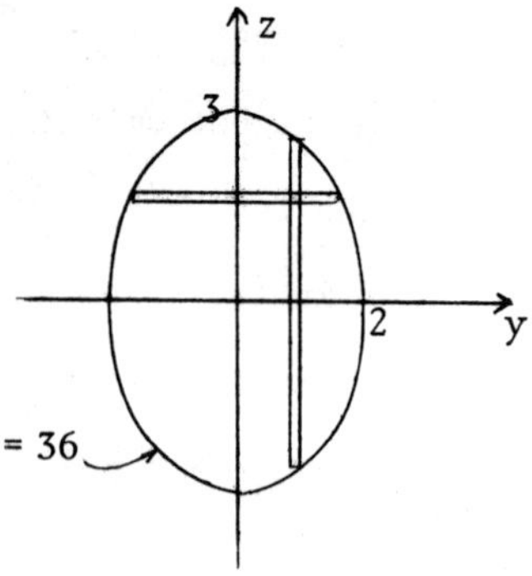

$$\int_{-2}^{2}\int_{-\frac{1}{2}\sqrt{36-9y^2}}^{\frac{1}{2}\sqrt{36-9y^2}}\int_{-\frac{1}{6}\sqrt{36-9y^2-4z^2}}^{\frac{1}{6}\sqrt{36-9y^2-4z^2}} f(x,y,z)\,dxdzdy$$

and $$\int_{-3}^{3}\int_{-\frac{1}{3}\sqrt{36-4z^2}}^{\frac{1}{3}\sqrt{36-4z^2}}\int_{-\frac{1}{6}\sqrt{36-9y^2-4z^2}}^{\frac{1}{6}\sqrt{36-9y^2-4z^2}} f(x,y,z)\,dxdydz.$$

11. $V = \int_{-2}^{2}\int_{0}^{4-x^2}\int_{0}^{4-z} dy\,dz\,dx$ or $\int_{0}^{4}\int_{0}^{4-y}\int_{-\sqrt{4-z}}^{\sqrt{4-z}} dx\,dz\,dy.$

13. The quickest way is to integrate with respect to x first. In Q, x varies from the back face $x = 0$, to the front face $x+z = 4$, or $x = 4-z$. The region R is in the yz-plane between the parabolas $y = z^2$ and $y = 2-z^2$. The points of intersection occur when $z^2 = 2-z^2$ or $z^2 = 1$ or $z = \pm 1$ and $R = \{(y,z): -1 \le z \le 1,\ z^2 \le y \le 2-z^2\}$.

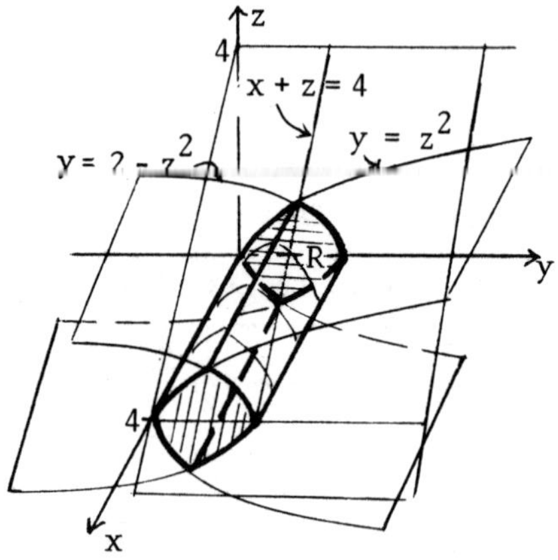

Then $V = \iiint_Q 1\,dV = \int_{-1}^{1}\int_{z^2}^{2-z^2}\int_{0}^{4-z} 1\,dxdydz$

$= \int_{-1}^{1}\int_{z^2}^{2-z^2} (4-z)\,dydx = \int_{-1}^{1} (4-z)(2-2z^2)\,dz$

$= 2\int_{-1}^{1} (4-z-4z^2+z^3)\,dz = \frac{32}{3}.$

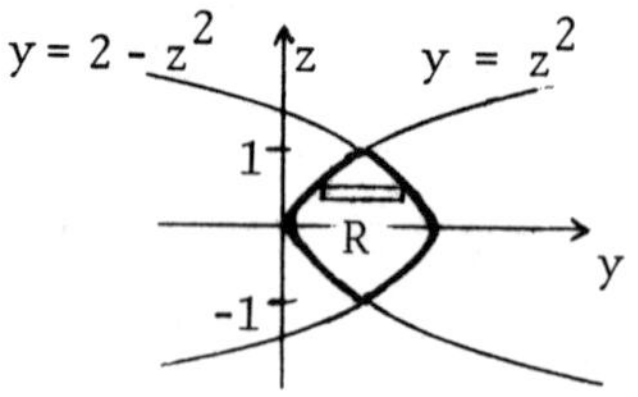

15. $V = \int_{-1}^{1}\int_{-\sqrt{1-y^2}}^{\sqrt{1-y^2}}\int_{0}^{2-y-z} dxdzdy.$

16. There are 2 choices. We could integrate z first from x^2+y^2 to $2-y$. The curve of intersection satisfies $x^2+y^2 = 2-y$ or $x^2+(y+\frac{1}{2})^2 = \frac{9}{4}$, which, when projected onto the xy-plane forms the boundary of the region R. A simpler region is obtained by integrating x 1st from the back to the front faces, $x = \mp\sqrt{z-y^2}$. In this case the region R is in the yx-plane bounded by the parabola $z = y^2$ and the line $z = 2-y$, and $R = \{(y,z)\ -2 \le y \le 1,\ y^2 \le z \le 2-y\}$.

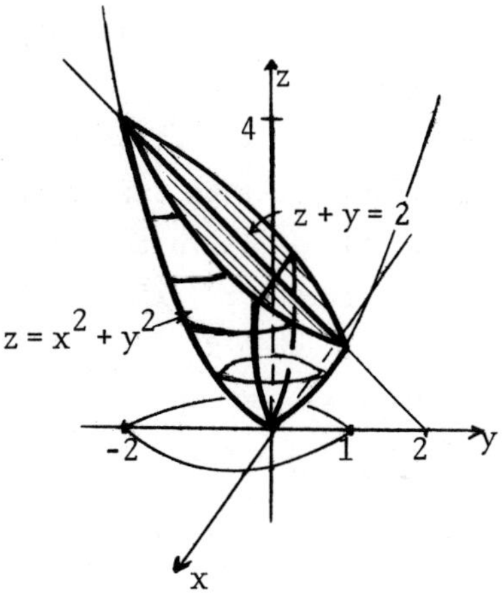

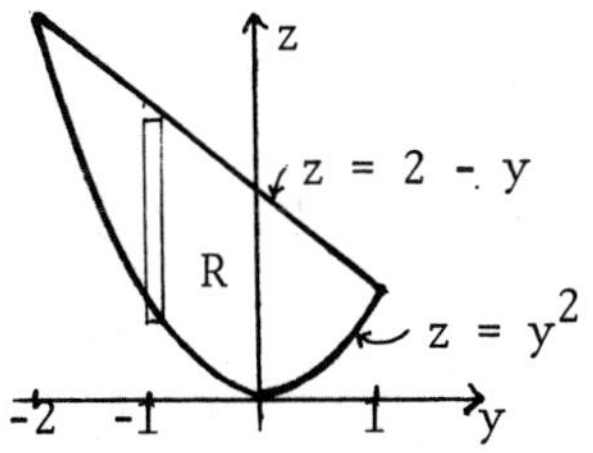

Then $V = \iiint_Q 1\ dV = \int_{-2}^{1}\int_{y^2}^{2-y}\int_{-\sqrt{z-y^2}}^{\sqrt{z-y^2}} dxdzdy$

$= 2\int_{-2}^{1}\int_{y^2}^{2-y} (z-y^2)^{1/2}\ dzdy =$

$2\int_{-2}^{1} (\frac{2}{3})(z-y^2)^{3/2}\Big]_{y^2}^{2-y}\ dy =$

$\frac{4}{3}\int_{-2}^{1} (2-y-y^2)^{3/2}\ dy = \frac{4}{3}\int_{-2}^{1} (\frac{9}{4} - (y+\frac{1}{2})^2)^{3/2} dy.$

Now let $u = y + 1/2$ and then $u = \frac{3}{2}\sin\theta$. The integrand is $(\frac{9}{4} - \frac{9}{4}\sin^2\theta)^{3/2}$ $= (\frac{3}{2})^3\cos^3\theta$ and $dy = du = \frac{3}{2}\cos\theta\ d\theta$. For the limits, $y = 1 \Rightarrow u = 3/2 \Rightarrow \theta = \pi/2$, $y = -2 \Rightarrow u = -3/2 \Rightarrow \theta = -\pi/2$, and, using $\cos^2 t = (1 + \cos 2t)/2$ several times we get

$$V = \frac{4}{3}(\frac{3}{2})^4\int_{-\frac{\pi}{2}}^{\frac{\pi}{2}} \cos^4\theta\ d\theta = \frac{27}{4}\int_{-\frac{\pi}{2}}^{\frac{\pi}{2}} (\frac{3}{8} + \frac{\cos 2\theta}{2} + \frac{\cos 4\theta}{8})d\theta = \frac{27}{4}\cdot\frac{3}{8}\cdot\pi = \frac{81\pi}{32}.$$

17. $V = \int_{-3}^{3}\int_{0}^{9-x^2}\int_{-1}^{2} dydzdx.$

19. The solid lies between the two parallel cylinders, $z = x^2$ and $z = x^3$ and extends from the xz-plane ($y = 0$) on the left to the cylinder $y = z^2$ on the right. Thus it is simplest to integrate y first (from 0 to z^2). R is in the xz-plane bounded by the curves $z = x^2$ and $z = x^3$ (which intersect when $x^2 = x^3 \Longrightarrow x^3 - x^2 = x^2(x-1) = 0 \Longrightarrow x = 0,1$). Thus $R = \{(x,z): 0 \le x \le 1, x^3 \le z \le x^2\}$ and

$$V = \int_0^1 \int_{x^3}^{x^2} \int_0^{z^2} dydzdx =$$

$$\int_0^1 \int_{x^3}^{x^2} z^2\, dzdx = \int_0^1 \frac{z^3}{3}\Big]_{x^3}^{x^2} dx = \frac{1}{3}\int_0^1 (x^6 - x^9)\,dx =$$

$$\frac{1}{3}\left(\frac{1}{7} - \frac{1}{10}\right) = \frac{1}{70}.$$

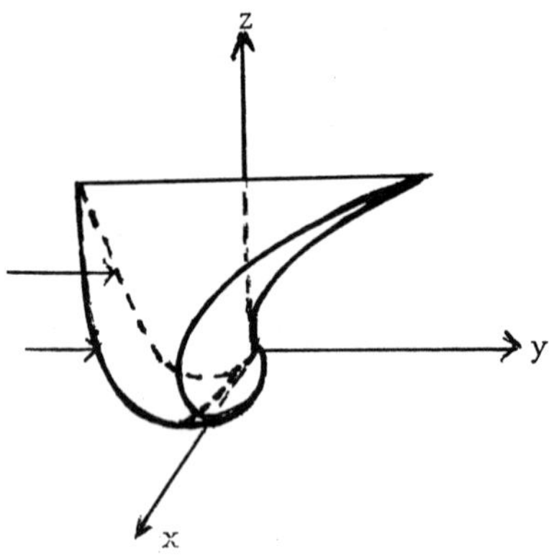

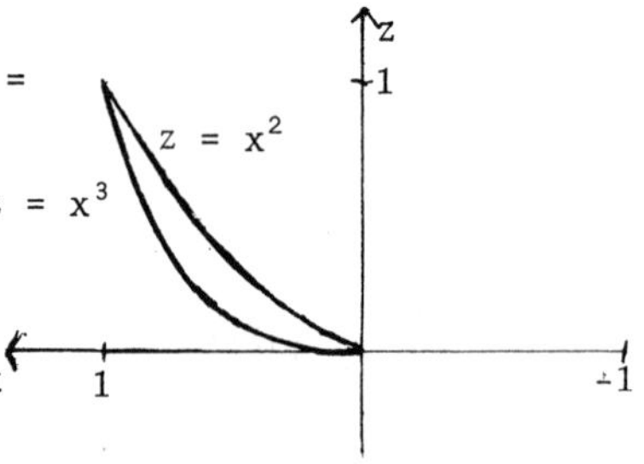

22. From the given y-limits of 0 and 4-x, the solid extends from the xz-plane ($y = 0$) to the plane $y = 4-x$. The region R is in the xz-plane and, from the given x and z limits, is of the form $R = \{(x,z): z^3 \le x \le \sqrt{z}, 0 \le z \le 1\}$, and, thus, is bounded by the curves $x = z^3$ and $x = \sqrt{z}$ (or $z = x^{1/3}$ and $z = x^2$) which intersect at the points (0,0) and (1,1). Thus Q is the region to the right of the region R, described above, and to the left of the plane $y = 4-x$.

25. From the given limits, y varies between $\mp\sqrt{z-x^2}$ so that the surface is $y^2 = z-x^2$ or $z = x^2+y^2$. R is in the xz-plane and is of the form $\{(x,z): 1 \le z \le 2, -\sqrt{z} \le x \le \sqrt{z}\}$ and thus is bounded by $z = x^2$, $z = 1$, $z = 2$. Trying these together leads to the given answer.

28. Hint: Mimic the process leading to (17.33) except the subregions are to be formed by means of "hyperplanes," (x,y,z,t = constants) in 4-dimensional space. The "volume" of a subregion Q_i is $\Delta x_i \Delta y_i \Delta z_i \Delta t_i$.

EXERCISES 17.7, page 870

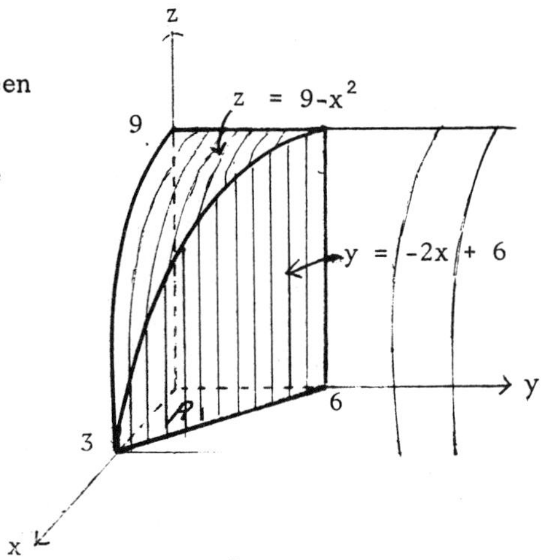

1. Choosing z to be integrated first, it is seen that z varies from the xy-plane (z = 0) to the cylinder $z = 9-x^2$. The region R in the xy-plane is bounded by the coordinate axes and the line $y = -2x+6$, i.e. $R = \{(x,y): 0 \le x \le 3,\ 0 \le y \le -2x+6\}$. Thus, taking the density to be 1,

$$M = V = \iiint_Q dV = \int_0^3 \int_0^{-2x+6} \int_0^{9-x^2} dz\,dy\,dx\ .$$

The answer in the text now follows from (17.41).

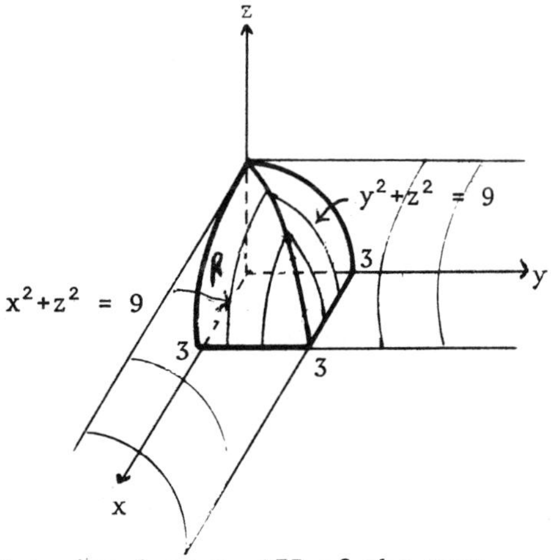

4. Choosing y to be integrated first, it is seen that y varies from the xz-plane (y = 0) to the cylinder $y^2+z^2 = 9$ which we have to rewrite as $y = \sqrt{9-z^2}$. The region R in the xz-plane is in the 1st quadrant portion of the circular region enclosed by $x^2+z^2 = 9$. Thus $R = \{(x,z): 0 \le x \le 3,\ 0 \le z \le \sqrt{9-x^2}\}$. Now let $I = \int_0^3 \int_0^{\sqrt{9-x^2}} \int_0^{\sqrt{9-z^2}} f(x,y,z)\ dy\,dz\,dx$ and use the method given in the answer to Exercise 1 on p. A75 of the text.

5. With the "fixed corner" at the origin and the cube in the 1st octant, $\rho(x,y,z) = k(x^2+y^2+z^2)$ and $M = \int_0^a \int_0^a \int_0^a \rho(x,y,z)\ dx\,dy\,dz$. Now use (17.41) and (17.42).

6. See #10 below for sketch and mass calculation.

7. We'll choose to integrate y first. In Q, y varies between the left and right faces, $y = \mp\sqrt{x-4z^2}$. R is the region in the xz-plane bounded by the parabola $x = 4z^2$ and $x = 4$. Thus $R = \{(x,z): 0 \le x \le 4, -\frac{\sqrt{x}}{2} \le z \le \frac{\sqrt{x}}{2}\}$, and with the density $\rho(x,y,z) = x^2+z^2$ we have

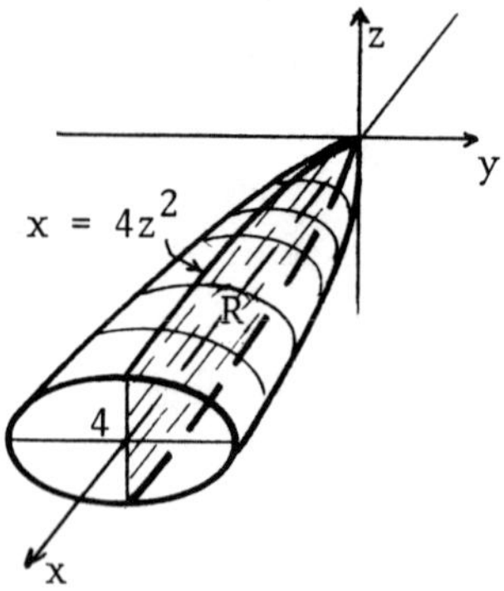

$$M = \iiint_Q \rho\, dV = \int_0^4 \int_{-\sqrt{x}/2}^{\sqrt{x}/2} \int_{-\sqrt{x-4z^2}}^{\sqrt{x-4z^2}} (x^2+z^2)\,dydzdx$$

and $$\bar{x} = \int_0^4 \int_{-\frac{\sqrt{x}}{2}}^{\frac{\sqrt{x}}{2}} \int_{-\sqrt{x-4z^2}}^{\sqrt{x-4z^2}} x(x^2+z^2)\,dydzdx \Big/ M.$$

The $\bar{y}$ and $\bar{z}$ integrals have integrands $y(x^2+z^2)$ and $z(x^2+z^2)$ respectively. Note that any other order of integration is just as easy to do.

9. Placing the cube as in #5 and selecting the edge along the z-axis, we have, with $\rho(x,y,z) = k$, $I_z = \int_0^a \int_0^a \int_0^a (x^2+y^2) \cdot k\, dxdydz$.

10. In Q, z ranges from 0 to $10-2x-5y = 2(5-x)-5y$, and $R = \{(x,y): 0 \le x \le 5, 0 \le y \le 2 - \frac{2}{5}x = \frac{2}{5}(5-x)\}$. The density, $\rho(x,y,z) = k|y|$, which is simply ky since $y \ge 0$ in Q. Thus M =

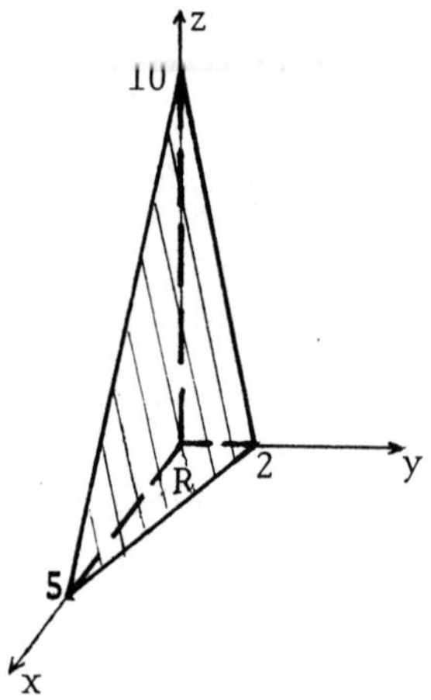

$$\iiint_Q \rho dV = \int_0^5 \int_0^{\frac{2}{5}(5-x)} \int_0^{2(5-x)-5y} ky\; dzdydx$$

$$= k\int_0^5 \int_0^{\frac{2}{5}(5-x)} (2(5-x)y - 5y^2)\; dydz$$

$$= k\int_0^5 (5-x)y^2 - \frac{5}{3}y^3\Big]_0^{\frac{2}{5}(5-x)}\; dx$$

$$= k\int_0^5 [(\tfrac{2}{5})^2(5-x)^3 - \tfrac{5}{3}(\tfrac{2}{5})^3(5-x)^3]\,dx$$

$$= k(\tfrac{4}{25} - \tfrac{8}{75}) \int_0^5 (5-x)^3 dx = k(\tfrac{1}{25} - \tfrac{2}{75})[-(5-x)^4]_0^5$$

$$= \frac{k}{75} \cdot 625 = \frac{25k}{3}.$$

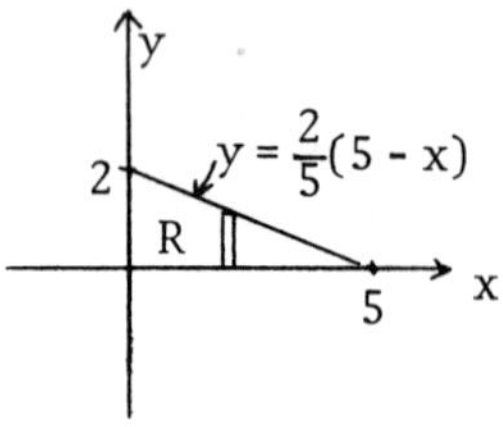

$$I_z = \iiint_Q (x^2+y^2)\rho\ dV = k\int_0^5 \int_0^{\frac{2}{5}(5-x)} \int_0^{2(5-x)-5y} (x^2+y^2)y\ dzdydx$$

$$= k\int_0^5 \int_0^{\frac{2}{5}(5-x)} [2x^2y(5-x)+2(5-x)y^3-5x^2y^2-5y^4]\,dydx$$

$$= k\int_0^5 [x^2(5-x)y^2 + \frac{1}{2}(5-x)y^4 - \frac{5}{3}x^2y^3 - y^5]_0^{\frac{2}{5}(5-x)}\ dx$$

$$= k\int_0^5 [(\frac{2}{5})^2x^2(5-x)^3 + \frac{1}{2}(\frac{2}{5})^4(5-x)^5 - \frac{5}{3}(\frac{2}{5})^3x^2(5-x)^3 - (\frac{2}{5})^5(5-x)^5]\,dx$$

$$= (\frac{2}{5})^2\ k\int_0^5 (5-x)^3[x^2 + \frac{2}{25}(5-x)^2 - \frac{2}{3}x^2 - (\frac{2}{5})^3(5-x)^2]\,dx$$

$$= (\frac{2}{5})^2\ k\int_0^5 (5-x)^3[\frac{x^2}{3} + (\frac{2}{25} - \frac{8}{125})(5-x)^2]\,dx\ .$$

When the integrand is multiplied out, the term $\int_0^5 (5-x)^5\ dx$ is easy, being equal to $-\frac{1}{6}(5-x)^6]_0^5 = -\frac{5^6}{6}$. Rather than expanding $(5-x)^3x^2$, it is much simpler to integrate by parts! With $u = x^2$, $dv = (5-x)^3dx$, $du = 2xdx$, $v = -\frac{(5-x)^4}{4}$ and $[uv]_0^5 = 0$. One more integration by parts yields $\int_0^5 (5-x)^5dx$ again. When all of these results are combined with the appropriate constants included, the result is $I_z = \frac{185}{9}k$. If d is the radius of gyration then $d^2 = I_z/M = \frac{185k}{9} \cdot \frac{3}{25k} = 37/15$ and $d = \sqrt{\frac{37}{15}}$.

13. The sphere has equation $x^2 + y^2 + z^2 = a^2$ and z varies between $\mp\sqrt{a^2-x^2-y^2}$. R is the region enclosed by the circle $x^2 + y^2 = a^2$. Using this data and $I_z = \iiint_Q (x^2+y^2)\rho(x,y,z)\,dV$, the answer given in the text results.

16. The answer is similar to that for #13 except that the z limits are $\pm c\sqrt{1 - \frac{x^2}{a^2} - \frac{y^2}{b^2}}$, and R is bounded by the ellipse $\frac{x^2}{a^2} + \frac{y^2}{b^2} = 1$. Thus R = $\{(x,y)\colon -a \le x \le a,\ -b\sqrt{1 - \frac{x^2}{a}} \le y \le b\sqrt{1 - \frac{x^2}{a}}\}$.

EXERCISES 17.8, page 876

1. In cylindrical coordinates, the equations are $z = r^2$, $r^2 = 4$, and $z = 0$. Within Q, z ranges from the xy-plane, $z = 0$, to the paraboloid $z = r^2$. The region R in the xy-plane is inside the entire circle $r = 2$ and $R = \{(r,\theta): 0 \le \theta \le 2\pi,\ 0 \le r \le 2\}$.

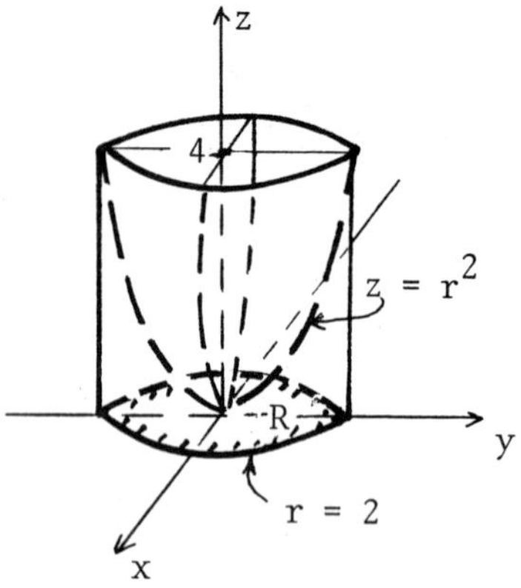

$$V = \iiint_Q dV = \int_0^{2\pi}\int_0^2\int_0^{r^2} r\,dz\,dr\,d\theta = \int_0^{2\pi}\int_0^2 r^3\,dr\,d\theta = 4\int_0^{2\pi} d\theta = 8\pi.$$

By symmetry of the solid about the z-axis, $\bar{x} = \bar{y} = 0$.

$$M_{xy} = \iiint_Q z\,dV = \int_0^{2\pi}\int_0^2\int_0^{r^2} zr\,dz\,dr\,d\theta = \int_0^{2\pi}\int_0^2 \frac{r^5}{2}\,dr\,d\theta = \frac{2^6}{12}\int_0^{2\pi} d\theta =$$

$$\frac{2^6\pi}{6} = \frac{32\pi}{3}, \text{ and } \bar{z} = \frac{M_{xy}}{V} = \frac{32\pi}{3}\cdot\frac{1}{8\pi} = \frac{4}{3}.$$

3. With the base on the xy-plane, the z-axis being the axis of the cylinder, and k = density, $I_z = \int_0^{2\pi}\int_0^a\int_0^h r^2 \cdot k \cdot r\,dzdrd\theta$ and $I_x = \int_0^{2\pi}\int_0^a\int_0^h (r^2\sin^2\theta + z^2) \cdot k \cdot r\,dzdrd\theta$.

4. The surfaces $z = r^2$ and $z = r$ intersect when $r^2 = r$, i.e. $r = 0$ and $r = 1$. Thus in Q, z varies from r^2 to r above R, the circular region of radius 1 centered at the origin. Since the solid is homogeneous, $\rho(x,y,z) = k$, constant, and $M = \int_0^{2\pi}\int_0^1\int_{r^2}^{r} kr\,dz\,dr\,d\theta =$

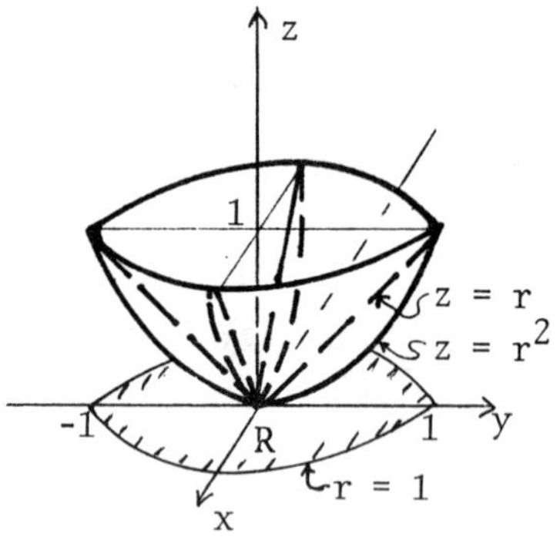

$$k\int_0^{2\pi}\int_0^1 (r^2 - r^3)\,dr\,d\theta = k\int_0^{2\pi}\left(\frac{1}{3} - \frac{1}{4}\right)d\theta = \frac{2\pi k}{12} = \frac{k\pi}{6}.$$

By homogeneity and symmetry, $\bar{x} = \bar{y} = 0$.

$$M_{xy} = \iiint_Q z\rho\,dV = k\int_0^{2\pi}\int_0^1\int_{r^2}^{r} zr\,dzdrd\theta = \frac{k}{2}\int_0^{2\pi}\int_0^1 (r^3 - r^5)\,drd\theta =$$

$$\frac{k}{2}\int_0^{2\pi}\left(\frac{1}{4} - \frac{1}{6}\right)d\theta = \frac{k\pi}{12} \text{ whence } \bar{z} = \frac{M_{xy}}{M} = \frac{k\pi}{12}\cdot\frac{6}{k\pi} = \frac{1}{2}.$$

$$I_z = \iiint_Q (x^2+y^2)\ \rho\ dV = k\int_0^{2\pi}\int_0^1\int_{r^2}^{r} r^2\cdot r\ dz\ dr\ d\theta =$$

$$k\int_0^{2\pi}\int_0^1 (r^4-r^5)\ drd\theta = k\int_0^{2\pi} (\tfrac{1}{5} - \tfrac{1}{6})\ d\theta = \frac{2\pi k}{30} = \frac{k\pi}{15}\ .$$

5. Placing the origin at the center of the sphere, its equation is $r^2+z^2 = a^2$

7. so that z ranges between $\pm\sqrt{a^2-r^2}$ and R is the circular region of radius a centered at the origin in the xy-plane. We'll take the line ℓ to be the z-axis. Then the distance of a point from ℓ is r and the density is $\rho(r,\theta,z) = kr$. Then

$$M = \iiint_Q \rho\ dV = \int_0^{2\pi}\int_0^a\int_{-\sqrt{a^2-r^2}}^{\sqrt{a^2-r^2}} (kr)r\ dz\ dr\ d\theta$$

$$= k\int_0^{2\pi}\int_0^a 2r^2\ \sqrt{a^2-r^2}\ dr\ d\theta = 4\pi k\int_0^a r^2\sqrt{a^2-r^2}\ dr.$$

To evaluate this, let $r = a \sin t$, so that $dr = a \cos t\ dt$, $\sqrt{a^2-r^2} = a \cos t$. When $r = 0$, $t = 0$ and then $r = a$, $t = \pi/2$. Thus

$$M = 4\pi k\int_0^{\pi/2} a^2\sin^2 t\cdot a^2\cos^2 t\ dt = \pi ka^4\int_0^{\pi/2} (1 - \cos 2t)(1 + \cos 2t)dt$$

$$= \pi ka^4\int_0^{\pi/2} (1 - \cos^2 2t)dt = \pi ka^4\int_0^{\pi/2} (\tfrac{1}{2} - \frac{\cos 4t}{2})dt = \frac{\pi^2 ka^4}{4}\ .$$

$$I_z = \iiint_Q r^2\ \rho\ dV = \int_0^a\int_0^{2\pi}\int_{-\sqrt{a^2-r^2}}^{\sqrt{a^2-r^2}} r^2(kr)r\ dr\ d\theta$$

$$= 2k\int_0^a\int_0^{2\pi} r^4\sqrt{a^2-r^2}\ d\theta dr = 4k\pi\int_0^a r^4\sqrt{a^2-r^2}\ dr.$$

With the same substitution as above, we get

$$I_z = 4\pi ka^6\int_0^{\pi/2} \sin^4 t\ \cos^2 t\ dt = 4\pi ka^6\int_0^{\pi/2} \frac{(1 - \cos 2t)^2}{4}\ \frac{(1 + \cos 2t)}{2}\ dt$$

$$= \frac{\pi ka^6}{2}\int_0^{\pi/2} (1 - \cos 2t - \cos^2 2t + \cos^3 2t)dt$$

$$= \frac{\pi ka^6}{2}\int_0^{\pi/2} (\tfrac{1}{2} - \cos 2t - \frac{\cos 4t}{2} + \cos^3 2t)dt$$

$$= \frac{\pi^2 ka^6}{8}$$ (The last 3 terms all yield 0 when integrated.)

If d is the radius of gyration then

$$d = \sqrt{\frac{I_z}{M}} = \sqrt{\frac{\pi^2 ka^6}{8}\cdot\ \frac{4}{\pi^2 ka^4}} = \frac{a}{\sqrt{2}}\ .$$

9. $M = \iiint_Q \text{density} \cdot dF = \int_0^{2\pi}\int_1^2\int_1^{5-r^2} (kz) \cdot r\, dz dr d\theta.$

10. In cylindrical coordinates the equations of the cylinder and sphere are $r = 2\sin\theta$ and $r^2 + z^2 = 4$, respectively. Within the solid, z ranges between $\mp\sqrt{4-r^2}$ and R is the circular region $\{(r,\theta): 0 \le \theta \le \pi,\ 0 \le r \le 2\sin\theta\}$. The density $\rho(r,\theta,z) = k|z|$. Since the solid is symmetric about the xy-plane, and also the density $(\rho(r,\theta,-z) = \rho(r,\theta,z))$, we'll calculate the mass of the top half (shown in sketch) where $0 \le z \le \sqrt{4-r^2}$, $\rho(r,\theta,z) = kz$, and double it.

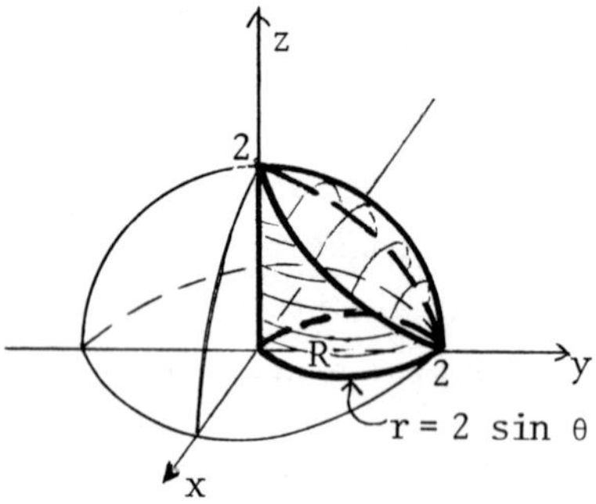

$$M = \iiint \rho\, dV = 2\int_0^{\pi}\int_0^{2\sin\theta}\int_0^{\sqrt{4-r^2}} (kz) r\, dz dr d\theta$$

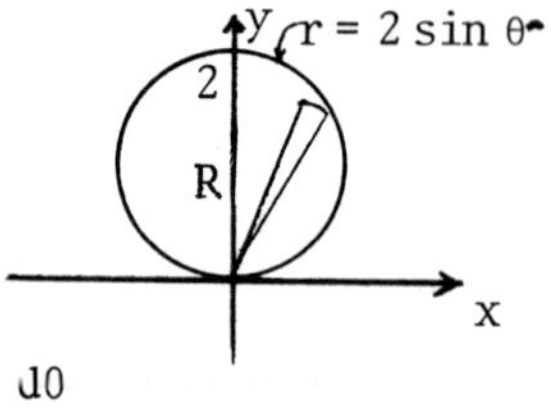

$$= 2k\int_0^{\pi}\int_0^{2\sin\theta} r\,\frac{z^2}{2}\Big]_0^{\sqrt{4-r^2}} dr\, d\theta$$

$$= k\int_0^{\pi}\int_0^{2\sin\theta} (4r - r^3)\, dr\, d\theta = k\int_0^{\pi} 2r^2 - \frac{r^4}{4}\Big]_0^{2\sin\theta} d\theta$$

$$= k\int_0^{\pi} (8\sin^2\theta - 4\sin^4\theta) d\theta = k\int_0^{\pi} [4 - 4\cos 2\theta - (1 - \cos 2\theta)^2]\, d\theta$$

$$= k\int_0^{\pi} [3 - 2\cos 2\theta - (\tfrac{1}{2} + \frac{\cos 4\theta}{2})] d\theta = \frac{5\pi}{2} k \text{ (The cosine terms all yield 0)}$$

For the center of mass, we first see that by the symmetries mentioned above, $\bar{z} = 0$, and from the symmetry about the yz-plane, $\bar{x} = 0$.

$$M_{xz} = \iiint y\rho\, dV = 2\int_0^{\pi}\int_0^{2\sin\theta}\int_0^{\sqrt{4-r^2}} (r\sin\theta)(kz) r\, dz dr d\theta$$

$$= k\int_0^{\pi}\int_0^{2\sin\theta} r^2 \sin\theta(4-r^2)\, dr\, d\theta = k\int_0^{\pi} \frac{4}{3} r^3 - \frac{1}{5} r^5\Big]_0^{2\sin\theta} \sin\theta\, d\theta$$

$$= k\int_0^{\pi} (\frac{32}{3}\sin^4\theta - \frac{32}{5}\sin^6\theta)\, d\theta = 32k\int_0^{\pi} [\frac{1}{3}(\frac{3}{8} - \frac{\cos 2\theta}{2} + \frac{\cos 4\theta}{8})$$

$$- \frac{1}{5}(\frac{5}{16} - \frac{3}{8}\cos 2\theta + \frac{3}{16}\cos 4\theta - \frac{\cos^3\theta}{8})] d\theta = 32k(\frac{\pi}{8} - \frac{\pi}{16}) = 2\pi k.$$

(Note: The usual identity $\sin^2 t = \dfrac{1 - \cos 2t}{2}$ was used several times above. The final cosine terms all yield 0.)

Finally, $\bar{y} = M_{xz}/M = 2\pi k \cdot \dfrac{2}{5\pi k} = \dfrac{4}{5}$ and the center of mass is (0,4/5,0).

11. See #13 below for set-up and notation.

$$M = \int_0^{\pi/2}\int_0^a\int_0^{2\pi} (k\rho)\cdot\rho^2 \sin\phi \, d\theta \, d\rho \, d\phi.$$

13. Placing the origin at the center of the sphere, its equation is $x^2 + y^2 + z^2 = a^2$, or $\rho = a$ in spherical coordinates. The z-axis is the axis of symmetry and so we want $I_z = \iiint_Q (x^2 + y^2)D\, dV$ where D is the density, Here $D(\rho,\theta,\phi) = k\rho$ where k is a constant, and ρ is the spherical coordinate--the distance of the point from the origin, which is the center of the base of the hemisphere. Since the hemisphere = $\{(\rho,\theta,\phi): 0 \le \rho \le a, 0 \le \theta \le 2\pi, 0 \le \phi \le \pi/2\}$ and $x^2 + y^2 = \rho^2\sin^2\phi$, we obtain

$$I_z = \int_0^{\pi/2}\int_0^a\int_0^{2\pi} (\rho^2\sin^2\phi)(k\rho)(\rho^2\sin\phi)d\theta d\rho d\phi = 2\pi k\int_0^{\pi/2}\int_0^a \rho^5 \sin^3\phi \, d\rho \, d\phi$$

$$= 2\pi k \frac{a^6}{6}\int_0^{\pi/2} \sin^3\phi \, d\phi = \frac{\pi k a^6}{3}\int_0^{\pi/2} (1 - \cos^2\phi)\sin\phi \, d\phi$$

$$= \frac{\pi k a^6}{3}\left[-\cos\phi + \frac{\cos^3\phi}{3}\right]_0^{\pi/2} = \frac{\pi k a^6}{3}\cdot\frac{2}{3}.$$

15. $$V = \iiint_Q 1 \, dV = \int_0^{\pi/4}\int_0^{2\pi}\int_0^{4\cos\phi} \rho^2\sin\phi \, d\rho d\theta d\phi$$

16. We'll calculate the volume of that portion of the solid in the 1st octant and multiply by 8. Within the solid $0 \le \rho \le 1$, $0 \le \theta \le \pi/2$, $\pi/4 \le \theta \le \pi/2$. (The cone $z^2 = x^2 + y^2$ makes an angle of $\pi/4$ with the z-axis and hence has spherical equation $\phi = \pi/4$. Of course, $\phi = \pi/2$ is the xy-plane.)

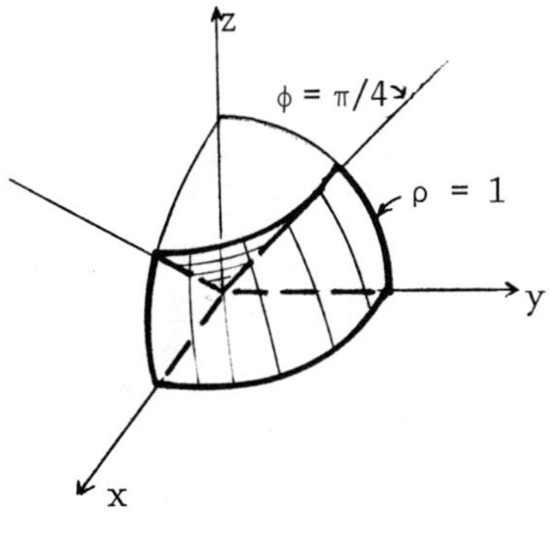

$$V = \iiint_Q 1 \, dV = 8\int_{\pi/4}^{\pi/2}\int_0^1\int_0^{\pi/2} \rho^2 \sin\phi \, d\theta d\rho d\phi$$

$$= 4\pi\int_{\pi/4}^{\pi/2}\int_0^1 \rho^2 \sin\phi \, d\rho \, d\phi$$

$$= \frac{4}{3}\pi\int_{\pi/4}^{\pi/2} \sin\phi \, d\phi = \frac{4}{3}\pi[-\cos\phi]_{\pi/4}^{\pi/2} = \frac{4\pi}{3\sqrt{2}}.$$

17. $$M = \int_0^{\pi}\int_0^{2\pi}\int_1^2 (k\rho^2)\cdot\rho^2 \sin\phi \, d\rho \, d\theta \, d\phi.$$

19. The thickness of the box is of course, $\Delta\rho_i$. One of the base dimensions is the length of a circular arc of radius ρ_i and central angle $\Delta\phi_i$. Recalling what radian measure is, the length of this arc is $\rho_i\Delta\phi_i$. The other base dimension is the length of a circular arc of radius $\rho_i \sin \phi_i$ (actually the polar coordinate, r) and central angle $\Delta\theta_i$. Combining these gives the result.

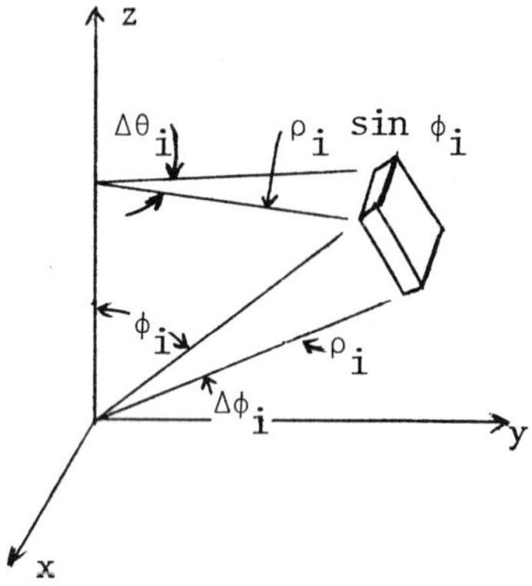

EXERCISES 17.9, page 880

1. $z = y + \frac{x^2}{2} = f(x,y) \Rightarrow f_x = x,\ f_y = 1 \Rightarrow \sqrt{f_x^2 + f_y^2 + 1} = \sqrt{x^2+2}$.

$$A = \int_0^1\int_0^1 \sqrt{x^2+2}\,dydx = \int_0^1 \sqrt{x^2+2}\,dx = \frac{1}{2}[x\sqrt{x^2+2} + 2\ln(x+\sqrt{x^2+2})]_0^1 =$$

$\frac{1}{2}[\sqrt{3} + 2\ln(1+\sqrt{3}) - 2\ln\sqrt{2}]$. (Formula 21 in the table of integrals was used for the last integral.)

3. $$A = \int_{-d}^{d}\int_{-\sqrt{d^2-x^2}}^{\sqrt{d^2-x^2}} \sqrt{(\frac{c}{a})^2 + (\frac{c}{b})^2 + 1}\,dy\,dx.$$

4. $z = \sqrt{4-x^2-y^2} \Rightarrow f_x = \frac{-x}{\sqrt{4-x^2-y^2}},\ f_y = \frac{-y}{\sqrt{4-x^2-y^2}} \Rightarrow f_x^2 + f_y^2 + 1 =$

$\frac{x^2}{4-x^2-y^2} + \frac{y^2}{4-x^2-y^2} + 1 = \frac{x^2+y^2+4-x^2-y^2}{4-x^2-y^2} \Rightarrow \sqrt{f_x^2 + f_y^2 + 1} = \frac{2}{\sqrt{4-x^2-y^2}}$

$$A = \int_{-1}^{1}\int_{-1}^{1} \frac{2}{\sqrt{4-x^2-y^2}}\,dy\,dx \approx 0.551 \quad \text{(using a computer)}$$

5. If R is the region inside the circle $x^2+y^2 = ay$, then the area of the portion of the surface above the xy-plane is $A = \iint_R \sqrt{1 + \frac{x^2}{a^2-x^2-y^2} + \frac{y^2}{a^2-x^2-y^2}}\,dA =$

$\int_0^{\pi}\int_0^{a\sin\theta} \sqrt{1 + \frac{r^2}{a^2-r^2}}\,r\,dr\,d\theta$. (Double this for the total surface area.)

7. The plane intersects the paraboloid in the curve $x^2+y^2 = 1$. Thus the region R above which the surface lies is inside the circle $x^2+y^2 = 1$. In polar coordinates $R = \{(r,\theta): 0 \le r \le 1,\ 0 \le \theta \le 2\pi\}$. $z = x^2+y^2 \Rightarrow f_x = 2x$, $f_y = 2y \Rightarrow \sqrt{f_x^2 + f_y^2 + 1} = \sqrt{4(x^2+y^2) + 1}$.

$$A = \iint_R \sqrt{4(x^2+y^2)+1}\ dA = \int_0^1 \int_0^{2\pi} \sqrt{4r\ +1}\ r\ d\theta\ dr = 2\pi\int_0^1 (4r^2+1)^{1/2}\ r\ dr =$$

$$2\pi(\tfrac{2}{3})(\tfrac{1}{8})(4r^2+1)^{3/2}\Big]_0^1 = \frac{\pi}{6}(5^{3/2}-1).$$

9. $$A = \int_{-2}^{2} \int_{-\sqrt{4-x^2}}^{\sqrt{4-x^2}} \sqrt{1+x^2+y^2}\ dy\ dx = \int_0^{2\pi} \int_0^2 \sqrt{1+r^2}\ r\ dr\ d\theta.$$

10. Since $z = x^2-y^2$, notice that $z < 0$ if $|y| > |x|$. Thus to get the surface area of the portion in the 1st octant, the region R must be below the line $y = x$ (or $\theta = \pi/4$). Moreover, since the portion inside $x^2+y^2 = 1$ (or $r = 1$) is desired $R = \{(r,\theta): 0 \le \theta \le \pi/4,\ 0 \le r \le 1\}$. The same integrand is obtained as in #7 above. Thus

$$A = \int_0^1 \int_0^{\pi/4} \sqrt{4r^2+1}\ r\ d\theta\ dr = \frac{\pi}{4}\int_0^1 \sqrt{4r^2+1}\ r\ dr = \frac{\pi}{4}\cdot\frac{2}{3}\cdot\frac{1}{8}(5^{3/2}-1) = \frac{\pi}{48}(5^{3/2}-1).$$

EXERCISES 17.10 (Review), page 882

1. $$\int_{-1}^{0} \int_{x+1}^{x^3} (x^2-2y)\ dydx = \int_{-1}^{0} x^2y - y^2\Big]_{x+1}^{x^3} dx = \int_{-1}^{0} [x^5-x^6-x^2(x+1)+(x+1)^2]dx =$$

$$\int_{-1}^{0} [x^5-x^6-x^3-x^2+(x+1)^2]dx = \frac{x^6}{6} - \frac{x^7}{7} - \frac{x^4}{4} - \frac{x^3}{3} + \frac{(x+1)^3}{3}\Big]_{-1}^{0} =$$

$$\frac{1}{3} - (\frac{1}{6} + \frac{1}{7} - \frac{1}{4} + \frac{1}{3}) = -\frac{5}{84}.$$

4. $$\int_2^0 \int_0^{z^2} \int_x^z (x+z)\ dydxdz = \int_2^0 \int_0^{z^2} (x+z)(z-x)\ dx\ dz = \int_2^0 \int_0^{z^2} (z^2-x^2)\ dx\ dz =$$

$$\int_2^0 z^2x - \frac{x^3}{3}\Big]_0^{z^2} dz = \int_2^0 (z^4 - \frac{z^6}{3})dz = \frac{z^5}{5} - \frac{z^7}{21}\Big]_2^0 = -\frac{32}{5} + \frac{128}{21} = -\frac{32}{105}.$$

7. R is the region enclosed by the right branch of the hyperbola $x^2-y^2 = 4$ and the line $x = 4$. Thus $R = \{(x,y): 2 \le x \le 4,\ -\sqrt{x^2-4} \le y \le \sqrt{x^2-4}\}$ yields the given answer.

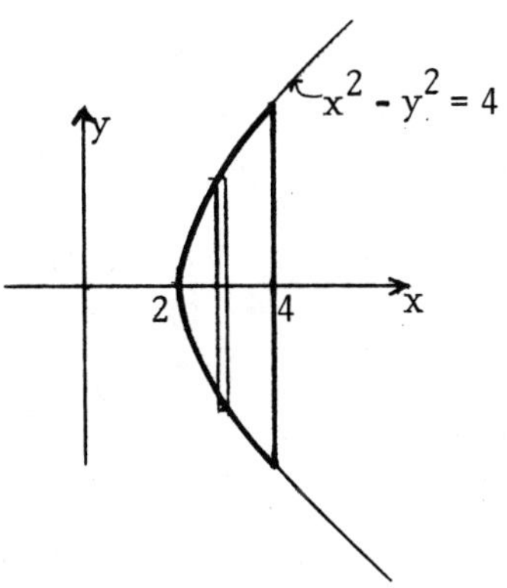

10. The curves intersect when $-x^2+4 = 3x^2$, i.e. at $x = \underline{+}1$. For such x, $3x^2 \leq y \leq -x^2+4$ in R and thus the integral is

$$\int_{-1}^{1}\int_{3x^2}^{-x^2+4} f(x,y)\,dydx.$$

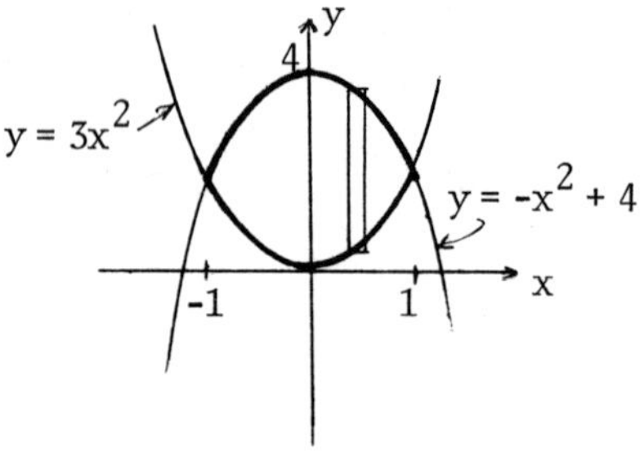

13. From the limits of integration, $R = \{(x,y): y^2 \leq x \leq 9,\ 0 \leq y \leq 3\}$, the region shown--a Type II region--bounded by $y = 0$, $x = 9$ and the upper half of the parabola $x = y^2$. It is also a Type I region with upper boundary the parabola, now written as $y = \sqrt{x}$. Thus

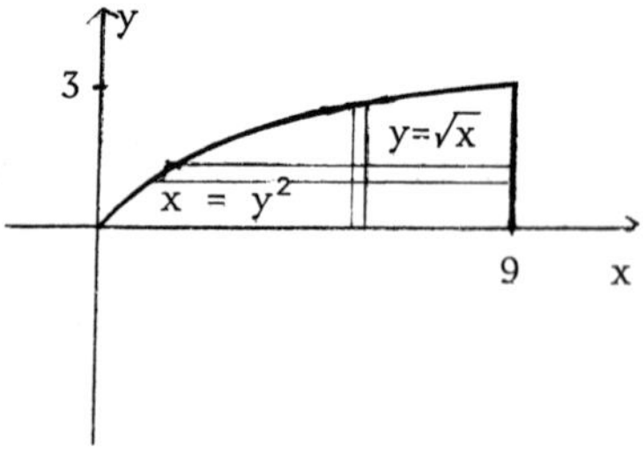

$$\int_0^3\int_{y^2}^{9} ye^{-x^2}\,dx\,dy = \int_0^9\int_0^{\sqrt{x}} ye^{-x^2}\,dy\,dx =$$

$$\int_0^9 \frac{y^2}{2}\Big]_0^{\sqrt{x}} e^{-x^2}\,dx = \frac{1}{2}\int_0^9 xe^{-x^2}\,dx = -\frac{1}{4}e^{-x^2}\Big]_0^9$$

$$= -\frac{1}{4}(e^{-81}-1) = \frac{1-e^{-81}}{4}.$$ (Note that with the original order of integration, the integral could not be done.)

15. $M = \int_0^3\int_x^{2x} (kx)\,dy\,dx.$

16. The integrals are a little easier if we describe R as $R = \{(x,y): -2 \leq y \leq 2,\ y^2 \leq x \leq 4\}$. The distance of a point (x,y) in R from the vertical line $x = -1$ is simply $x+1$. Thus $\rho(x,y) = k(x+1)$, k a constant, and

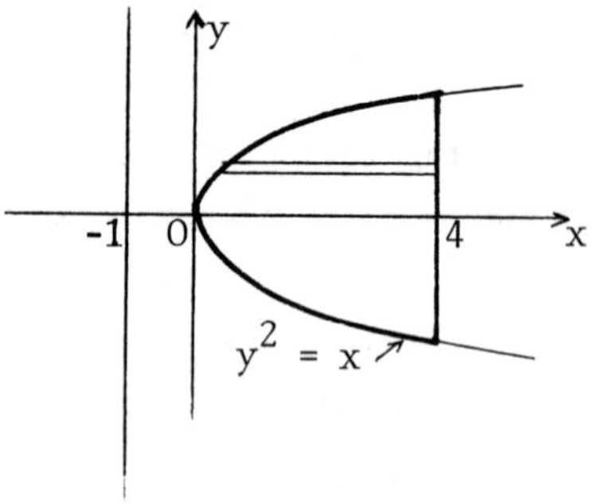

$$M = \iint_R \rho\,dA = \int_{-2}^{2}\int_{y^2}^{4} k(x+1)\,dx\,dy =$$

$$k\int_{-2}^{2} \frac{x^2}{2} + x\Big]_{y^2}^{4}\,dy =$$

$$k\int_{-2}^{2}\left(12 - \frac{y^4}{2} - y^2\right)dy = k\left[12y - \frac{y^5}{10} - \frac{y^3}{3}\right]_{-2}^{2} = 2k\left[24 - \frac{32}{10} - \frac{8}{3}\right] = \frac{544k}{15}.$$

The symmetry of R about the x-axis and independence of ρ on $y \Rightarrow \bar{y} = 0$. For $\bar{x}$, we compute

$$M_y = \iint_R x\,\rho\,dA = \int_{-2}^{2}\int_{y^2}^{4} x\cdot k(x+1)\,dx\,dy = k\int_{-2}^{2}\left[\frac{x^3}{3}+\frac{x^2}{2}\right]_{y^2}^{4} dy =$$

$$k\int_{-2}^{2}\left[\left(\frac{64}{3}+8\right)-\frac{y^6}{3}-\frac{y^4}{2}\right]dy = k\left[\frac{88}{3}y-\frac{y^7}{21}-\frac{y^5}{10}\right]_{-2}^{2} = 2k\left[\frac{176}{3}-\frac{128}{21}-\frac{32}{10}\right] =$$

$\frac{(32)(108)}{35}k$ and

$$\bar{x} = \frac{M_y}{M} = \frac{(32)(108)k}{35}\cdot\frac{15}{544k} = \frac{324}{119} \approx 2.7.$$

17. $M = \int_0^{2\pi}\int_0^{2+\sin\theta} \frac{k}{r}\cdot r\,dr\,d\theta.$

19. From the y-limits we see that the lower boundary of R is the curve $y = -\sqrt{a^2-x^2}$, the bottom half of the circle $x^2+y^2 = a^2$, and the upper boundary is the x-axis $(y = 0)$. Since x varies from $-a$ to 0 (so that $-a \le x \le 0$) the region lies in the 3rd quadrant bounded by the coordinate axes and the circle mentioned above. For points within this region, R, the polar coordinate, r, varies from 0 to a, and θ varies from π to $3\pi/2$. Recalling that $\sqrt{x^2+y^2} = r$ and $dA = r\,dr\,d\theta$ in polar form we obtain:

$$\int_{-a}^{0}\int_{-\sqrt{a^2-x^2}}^{0}\sqrt{x^2+y^2}\,dy\,dx = \iint_R \sqrt{x^2+y^2}\,dA = \int_{\pi}^{3\pi/2}\int_0^a r\cdot r\,dr\,d\theta =$$

$$\int_{\pi}^{3\pi/2}\left.\frac{r^3}{3}\right]_0^a d\theta = \frac{a^3}{3}\int_{\pi}^{3\pi/2} d\theta = \left.\frac{a^3}{3}\theta\right]_{\pi}^{3\pi/2} = \frac{a^3}{3}\left(\frac{3\pi}{2}-\pi\right) = \frac{a^3}{3}\cdot\frac{\pi}{2}.$$

21. With the right angle vertex at (0,0) and the side of length a along the y-axis we have $I_y = \int_0^b\int_0^{-ax/b+a} x^2\cdot k\cdot dy\,dx.$ (k = density.)

22. Placing the origin at the common center of the circles, $R = \{(r,\theta): a \le r \le b,\ 0 \le \theta \le 2\pi\}$ and $\rho(r,\theta) = kr$. Selecting the x-axis as the line through the center, then I_x is desired.

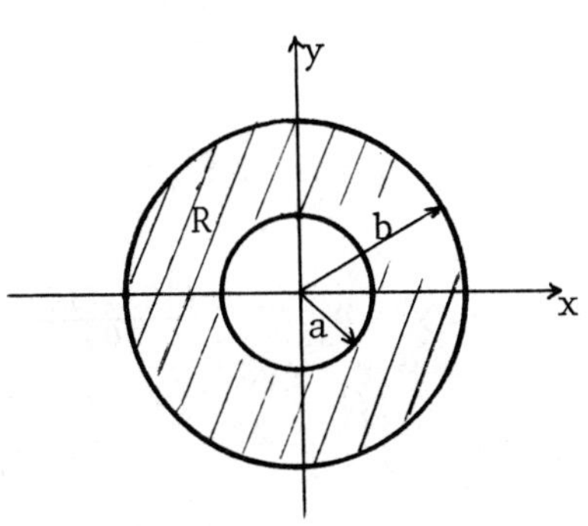

$$I_x = \iint_R y^2\,\rho\,dA = \int_0^{2\pi}\int_a^b (r\sin\theta)^2\cdot kr\cdot r\,dr\,d\theta$$

$$= k\int_0^{2\pi}\int_a^b r^4\sin^2\theta\,dr\,d\theta =$$

$$k\frac{(b^5-a^5)}{5}\int_0^{2\pi}\sin^2\theta\,d\theta$$

$$= \frac{k(b^5-a^5)}{5}\int_0^{2\pi}\left(\frac{1}{2}-\frac{\cos 2\theta}{2}\right)d\theta = \frac{k(b^5-a^5)}{5}\cdot\pi.$$

23. $V = \int_1^2 \int_1^3 xy^2 \, dy \, dx.$

25. The easiest way is to integrate 1st with respect to y. In Q, y ranges from the left face, the xz-plane, y = 0, to the slant face, the plane y+z = 4. Then R is the region in the xz-plane bounded by the parabola $z = x^2$ and the line z = 4. Thus $R = \{(x,z): -2 \le x \le 2, x^2 \le z \le 4\}$ and

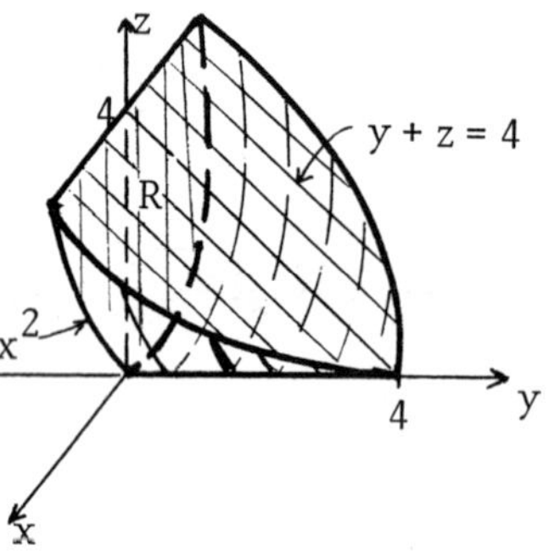

$$V = \iiint_Q 1 \, dV = \int_{-2}^2 \int_{x^2}^4 \int_0^{4-z} dy \, dz \, dx =$$

$$\int_{-2}^2 \int_{x^2}^4 (4-x) \, dz \, dx =$$

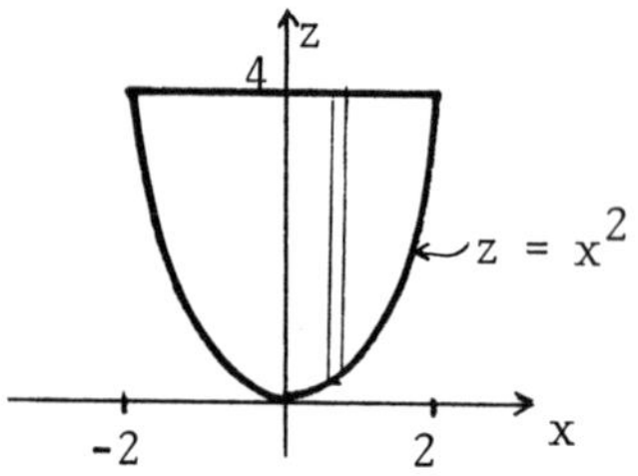

$$\int_{-2}^2 \left[-\frac{(4-z)^2}{2}\right]_{x^2}^4 dx = \int_{-2}^2 \frac{(4-x^2)^2}{2} dx$$

$$= \int_{-2}^2 (8 - 4x^2 + \frac{x^4}{2}) dx = 8x - \frac{4}{3}x^3 + \frac{x^5}{10}\Big]_{-2}^2$$

$$= 2(16 - \frac{32}{3} + \frac{32}{10}) = \frac{256}{15}.$$

For the centroid, we see that by symmetry of Q about the yz-plane, $\bar{x} = 0$. For $\bar{y}$ and $\bar{z}$, we proceed:

$$M_{xz} = \iiint_Q y \, dV = \int_{-2}^2 \int_{x^2}^4 \int_0^{4-z} y \, dy \, dz \, dx = \frac{1}{2}\int_{-2}^2 \int_{x^2}^4 (4-z)^2 \, dz \, dx$$

$$= \frac{1}{2}\int_{-2}^2 \left[-\frac{(4-z)^3}{3}\right]_{x^2}^4 dx = \frac{1}{6}\int_{-2}^2 (4-x^2)^3 \, dx = \frac{1}{6}\int_{-2}^2 (64-48x^2+12x^4-x^6) \, dx$$

$$= \frac{1}{6}[64x-16x^3 + \frac{12}{5}x^5 - \frac{x^7}{7}]_{-2}^2 = \frac{2}{6}(128 - 128 + \frac{12(32)}{5} - \frac{128}{7}) = \frac{(128)(16)}{105}$$

and $\bar{y} = M_{xz}/V = \frac{(128)(16)}{105} \cdot \frac{15}{256} = \frac{8}{7}$. Finally,

$$M_{xy} = \iiint_Q z \, dV = \int_{-2}^2 \int_{x^2}^4 \int_0^{4-z} z \, dy \, dz \, dx = \int_{-2}^2 \int_{x^2}^4 (4z-z^2) \, dz \, dx$$

$$= \int_{-2}^2 2z^2 - \frac{z^3}{3}\Big]_{x^2}^4 dx = \int_{-2}^2 (\frac{32}{3} - 2x^4 + \frac{x^6}{3}) \, dx = \frac{32}{3}x - \frac{2}{5}x^5 + \frac{x^7}{21}\Big]_{-2}^2$$

$$= 2(\frac{64}{3} - \frac{64}{5} + \frac{128}{21}) = \frac{1024}{35}$$

and $\bar{z} = \frac{M_{xy}}{V} = \frac{1024}{35} \cdot \frac{15}{256} = \frac{12}{7}$ and the centroid is $(0, \frac{8}{7}, \frac{12}{7})$.

28. Q is the solid between the cylinder $x^2+y^2 = 4$ (or $r = 2$) and the paraboloid $z = 9-x^2-y^2$ (or $z = 9-r^2$). Thus within Q, z varies from 0 to $9-r^2$. R is the region in the xy-plane between the circles $r = 2$ (the trace of the cylinder) and $r = 3$ (the trace of the paraboloid), and $R = \{(r,\theta): 2 \le r \le 3, 0 \le \theta \le 2\pi\}$. With $\rho(r,\theta,z) = k$, constant, we have

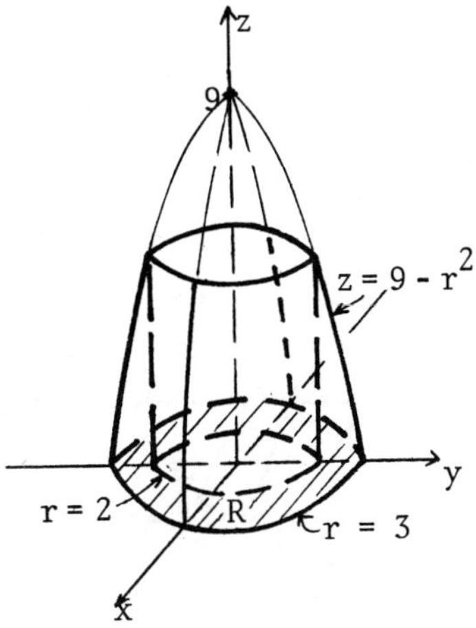

(a) $M = \iiint_Q \rho\, dV = k\int_2^3 \int_0^{2\pi} \int_0^{9-r^2} k\, r\, dz\, d\theta\, dr$

$= k\int_2^3 \int_0^{2\pi} (9r-r^3)d\theta\, dr = 2\pi k\int_2^3 (9r-r^3)dr$

$= 2\pi k[\frac{9}{2}r^2 - \frac{r^4}{4}]_2^3 = 2\pi k[\frac{81}{2} - \frac{81}{4} - 18 + \frac{16}{4}] = \frac{25\pi k}{2}$.

(b) For the center of mass, we have $\overline{y} = \overline{x} = 0$ by homogeneity and symmetry about the z-axis. For $\overline{z}$,

$M_{xy} = \iiint_Q z\rho\, dV = k\int_2^3 \int_0^{2\pi} \int_0^{9-r^2} z\cdot r\, dz\, dr\, d\theta = \frac{k}{2}\int_2^3 \int_0^{2\pi} (9-r^2)^2 r\, d\theta\, dr =$

$\pi k\int_2^3 (9-r^2)^2\, r\, dr = \pi k[-\frac{(9-r^2)^3}{6}]_2^3 = \pi k \cdot \frac{125}{6}$.

(c) $I_z = \iiint_Q (x^2+y^2)\rho\, dV = k\int_2^3 \int_0^{2\pi} \int_0^{9-r^2} r^2\cdot r\, dz\, d\theta\, dr$

$= k\int_2^3 \int_0^{2\pi} (9r^3-r^5)d\theta\, dr = 2\pi k\int_2^3 (9r^3-r^5)dr = 2\pi k[\frac{9}{4}r^4 - \frac{r^6}{6}]_2^3$

$= 2\pi k[\frac{729}{4} - \frac{729}{6} - 36 + \frac{64}{6}] = \frac{425\pi k}{6}$.

29. With the center at the origin, $M = \int_0^{\pi} \int_0^{2\pi} \int_0^a (k\rho) \cdot \rho^2 \sin\phi\, d\rho\, d\theta\, d\phi$.

CHAPTER 18

TOPICS IN VECTOR CALCULUS

EXERCISES 18.1, page 888

4.

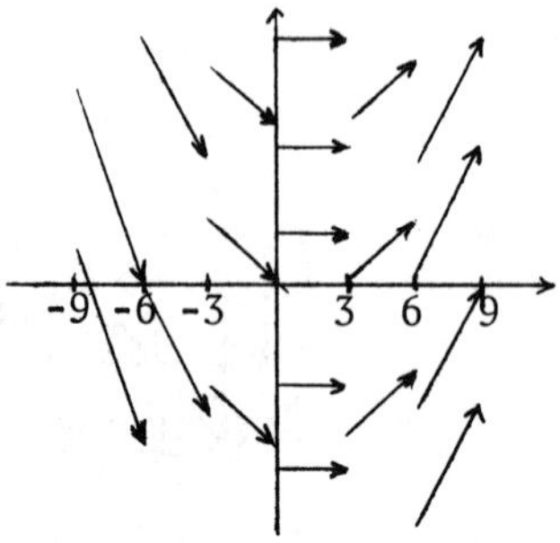

10. At every point (x,y,z), the vector points straight up 2 units in length.

13. The answer is simply $\underline{F} = \nabla f = f_x\underline{i} + f_y\underline{j} + f_z\underline{k} = 2x\underline{i} - 6y\underline{j} + 8z\underline{k}$.

16. Noting that $\frac{\partial}{\partial x}\ln(xyz) = \frac{yz}{xyz} = \frac{1}{x}$, etc., we obtain $\nabla f = \frac{1}{x}\underline{i} + \frac{1}{y}\underline{j} + \frac{1}{z}\underline{k}$.

19. $\nabla f = x\underline{i} + y\underline{j} + z\underline{k}$.

22. $\underline{F}$ is continuous at $P_0(x_0,y_0,z_0)$ if $\lim\limits_{(x,y,z)\to(x_0,y_0,z_0)} \underline{F}(x,y,z) = \underline{F}(x_0,y_0,z_0)$

which happens if and only if the component functions of $\underline{F}$, M, N and P, are continuous at P_0. Geometrically, if $\underline{F}$ is continuous at P_0 then vectors $\underline{F}(x,y,z)$ for points (x,y,z) near P_0 will have almost the same direction and magnitude as $\underline{F}(P_0)$, i.e. there will be no abrupt changes of either direction or magnitude.

EXERCISES 18.2, page 900

1. $x' = 3,\ y' = 3t^2 \Longrightarrow ds = \sqrt{9+9t^4}\,dt = 3\sqrt{1+t^4}\,dt,\ 0 \le t \le 1.$

$$\int_C (x^3+y)\,ds = \int_0^1 [(3t)^3+t^3]\cdot 3\sqrt{1+t^4}\,dt = \int_0^1 (28t^3)\cdot 3\sqrt{1+t^4}\,dt = 21\int_1^2 u^{1/2}du =$$

$$21\left(\tfrac{2}{3}\right)u^{3/2}\Big]_1^2 = 14(2^{3/2} - 1).$$ (The substitution of $u = 1+t^4$, $du = 4t^3dt$ was used above.) In the other 2 integrals recall $dx = x'dt = 3dt$, $dy = y'dt = 3t^2dt$.

$$\int_C (x^3+y)\,dx = \int_0^1 (28t^3)\cdot 3\,dt = 84\,\frac{t^4}{4}\Big]_0^1 = 21.$$

$$\int_C (x^3+y)\,dy = \int_0^1 (28t^3)\cdot 3t^2\,dt = 84\,\frac{t^6}{6}\Big]_0^1 = 14.$$

4. With initial point (0,0) and terminal point (2,8) we have x varying from 0 to 2. Also, $y = x^2+2x \Longrightarrow dy = (2x+2)dx$.

$$\int_C y\,dx + (x+y)\,dy = \int_0^2 (x^2+2x)\,dx + (x+(x^2+2x))(2x+2)\,dx = \int_0^2 [(x^2+2x) + (x^2+3x)(2x+2)]\,dx = \int_0^2 (2x^3+9x^2+8x)\,dx = \frac{x^4}{2} + 3x^3 + 4x^2\Big]_0^2 = 8 + 3\cdot 8 + 4\cdot 4 = 48.$$

7. (a) Let C be broken up into the paths C_1 and C_2 as shown. Then C_1: $x = t$, $y = 0 \Longrightarrow dx = dt$, $dy = 0\,dt$, $0 \le t \le 1$ and C_2: $x = 1$, $y = t \Longrightarrow dx = 0\,dt$, $dy = dt$, $0 \le t \le 3$. Then

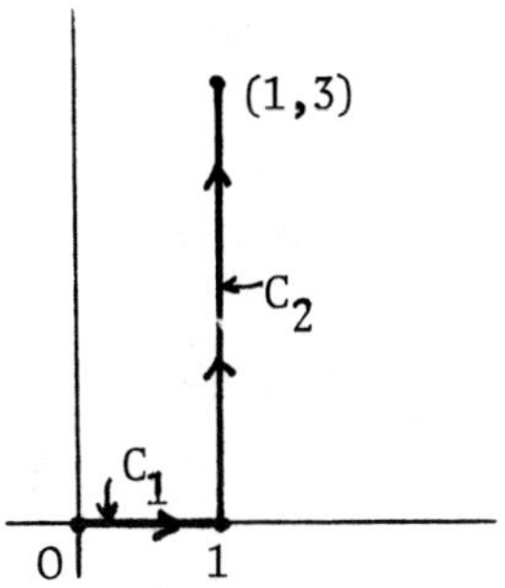

$$\int_{C_1} xy\,dx + (x+y)\,dy = \int_0^1 [0 + (t+0)\cdot 0]\,dt = 0,$$

$$\int_{C_2} xy\,dx + (x+y)\,dy = \int_0^3 [(1\cdot t)\cdot 0 + (1+t)]\,dt =$$

$$\int_0^3 (1+t)\,dt = \frac{(1+t)^2}{2}\Big]_0^3 = 8 - \frac{1}{2} = \frac{15}{2}$$ and the total integral along C is $\int_{C_1} + \int_{C_2} = \frac{15}{2}$.

(b) Now C is broken up differently as shown. Then C_1: $x = 0$, $y = t \Longrightarrow dx = 0\,dt$, $dy = 1\,dt$; $0 \le t \le 3$ and C_2: $x = t$, $y = 3 \Longrightarrow dx = dt$, $dy = 0\,dt$, $0 \le t \le 1$. Then $\int_{C_1} xy\,dx + (x+y)\,dy = \int_0^3 [0+(0+t)]\,dt = \frac{t^2}{2}\Big]_0^3 = \frac{9}{2}$, and $\int_{C_2} xy\,dx+(x+y)\,dy = \int_0^1 [3t+(t+3)\cdot 0]\,dt = \frac{3t^2}{2}\Big]_0^1 = \frac{3}{2}$. The integral along C is thus $\frac{9}{2} + \frac{3}{2} = 6$.

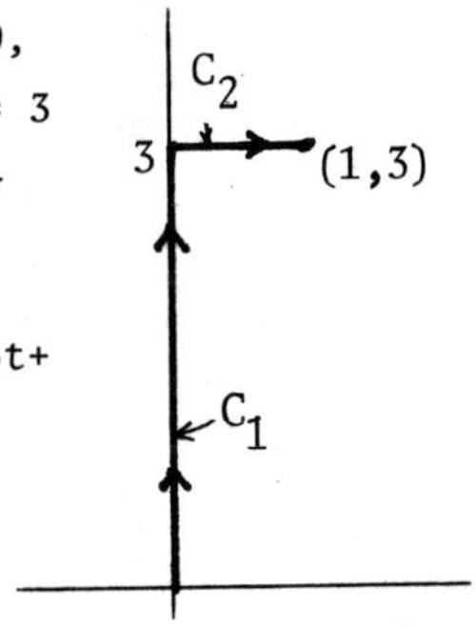

(c) C now has equation $y = 3x$, so that $dy = 3\,dx$, $0 \le x \le 1$.

$$\int_C xy\,dx + (x+y)\,dy = \int_0^1 [x(3x) + (x+3x)\cdot 3]\,dx = \int_0^1 (3x^2+12x)\,dx = x^3+6x^2\Big]_0^1 = 1 + 6 = 7.$$

(d) $y = 3x^2 \Longrightarrow dy = 6x\,dx$, $0 \le x \le 1$ and $\int_C xy\,dx + (x+y)\,dy = \int_0^1 [x(3x^2) + (x+3x^2)6x]\,dx = \int_0^1 (21x^3+6x^2)\,dx = \frac{21}{4}x^4 + 2x^3\Big]_0^1 = \frac{21}{4} + 2 = \frac{29}{4}$.

10. $x = \sin t$, $y = 2 \sin t$, $z = \sin^2 t \Rightarrow dx = \cos t\, dt$, $dy = 2 \cos t\, dt$, $dz = 2 \sin t \cos t\, dt$ for $0 \leq t \leq \pi/2$ and $\int_C y\, dx + x\, dy + x\, dz = \int_0^{\pi/2} [2 \sin t \cos t + \sin^2 t(2 \cos t) + \sin t(2 \sin t \cos t)]dt = \int_0^{\pi/2} (2 \sin t + 4 \sin^2 t)\cos t\, dt = \sin^2 t + \frac{4}{3} \sin^3 t\Big]_0^{\pi/2} = 1 + \frac{4}{3} = \frac{7}{3}$.

13. Parametric equations for the line segment from (0,0,0) to (1,2,3) are $x = t$, $y = 2t$, $z = 3t$, $0 \leq t \leq 1$, and $x' = 1$, $y' = 2$, $z' = 3 \Rightarrow ds = \sqrt{1^2+2^2+3^2}\, dt = \sqrt{14}\, dt$. $\int_C xyz\, ds = \int_0^1 t(2t)(3t)\sqrt{14}dt = 6\sqrt{14} \int_0^1 t^3 dt = \frac{3}{2}\sqrt{14}$.

15. Using $x = t$, $y = 4-t^2$, $-2 \leq t \leq 2$, $M = \int_{-2}^{2} k|t|\sqrt{1+(-2t)^2}\, dt$. Now use (18.8) and (18.9).

16. The shape of the wire is the upper half of the circle $x^2+y^2 = 1$. Parametric equations are, thus, $x = \cos t$, $y = \sin t$, $0 \leq t \leq \pi$ so that $x' = -\sin t$, $y' = \cos t$, $ds = \sqrt{(-\sin t)^2 + (\cos t)^2}\, dt = 1\, dt$. The density $\rho(x,y) = ky$ since y is the distance of a point (x,y) on the wire from the x-axis. Thus

$$M = \int_C \rho\, ds = \int_C ky\, ds = \int_0^{\pi} k \sin t\, dt = k \cos t\Big]_0^{\pi} = 2k.$$

$$M_y = \int_C x\rho\, ds = k\int_C xy\, ds = k\int_0^{\pi} \cos t \sin t\, dt = 0, \text{ so } \bar{x} = 0.$$

$$M_x = \int_C y\rho\, ds = k\int_C y^2\, ds = k\int_0^{\pi} \sin^2 t\, dt = \frac{k}{2}\int_0^{\pi} (1 - \cos 2t)dt = \frac{k\pi}{2} \text{ so that}$$

$$\bar{y} = \frac{M_x}{M} = \frac{k\pi}{2} \cdot \frac{1}{2k} = \frac{\pi}{4}.$$

19. Refer to #7 above for sketches and parametric equations. In each part of this problem, $\int_C \underline{F}\cdot d\underline{r} = \int_C xy^2 dx + x^2 y dy$.

(a) $\int_{C_1} xy^2 dx + x^2 y dy = \int_0^1 (t\cdot 0^2 + t^2\cdot 0)dt = 0$,

$\int_{C_2} xy^2 dx + x^2 y dy = \int_0^3 (t^2\cdot 0 + 1\cdot t)dt = \int_0^3 t\, dt = \frac{9}{2}$, and

$$\int_C \underline{F}\cdot d\underline{r} = 0 + \frac{9}{2}.$$

(b) $\int_{C_1} xy^2dx + x^2ydy = \int_0^3 0\ dt = 0$, $\int_{C_2} xy^2dx + x^2ydy = \int_0^1 9t\ dt = \frac{9}{2}$ and

$$\int_C \underline{F}\cdot d\underline{r} = \frac{9}{2} \text{ again.}$$

(c) $\int_C xy^2dx + x^2ydy = \int_0^1 [x(3x)^2 + x^2(3x)(3)]dx = \int_0^1 18x^3dx = \frac{9}{2}$.

(d) $\int_C xy^2dx + x^2ydy = \int_0^1 [x(3x^2)^2 + x^2(3x^2)(6x)]dx = \int_0^1 27x^5dx = \frac{9}{2}$.

(All answers are the same here, but not in #7. The reason will be apparent in the next section.)

21. Using $x = a\cos t$, $y = a\sin t$ with t varying from π to 0, $W = \int_C \underline{F}\cdot d\underline{r} =$

$$\int_\pi^0 \left[\frac{4\cos t}{a^2}(-a\sin t) + \frac{4\sin t}{a^2}(a\cos t)\right]dt.$$

22. Work $= W = \int_C \underline{F}\cdot d\underline{r} = \int_C (x^2+y^2)dx + xy\ dy$ where C is the path $y = x^2$, $0 \le x \le 2$,

so that $dy = 2x\ dx$.

$$W = \int_0^2 [x^2+x^4+x^3(2x)]dx = \int_0^2 (x^2+3x^4)dx = \frac{x^3}{3} + \frac{3}{5}x^5\Big]_0^2 = \frac{8}{3} + \frac{96}{5} = \frac{328}{15}.$$

23. $W = \int_0^2 (t^2\cdot 1 + t^3\cdot 2t + t\cdot 3t^2)dt.$

$t^2 + 2t^4 + 3t$

25. Refer to #16 for the parametric equations and density.

$$I_x = \int_C y^2\rho\ ds = k\int_C y^3 ds = k\int_0^\pi \sin^3 t\ dt = k\left[-\cos t + \frac{\cos^3 t}{3}\right]_0^\pi = \frac{4k}{3}.$$

$$I_y = \int_C x^2\rho\ ds = k\int_C x^2y\ ds = k\int_0^\pi \cos^2 t\sin t\ dt = -k\frac{\cos^3 t}{3}\Big]_0^\pi = \frac{2k}{3}.$$

28. $x' = -a\sin t$, $y' = a\cos t$, $z' = b \Rightarrow ds = \sqrt{a^2(\sin^2 t + \cos^2 t) + b^2}\ dt = \sqrt{a^2+b^2}\ dt$. The density $\rho = k$, constant.

$$I_z = \int_C (x^2+y^2)\rho\ ds = \int_0^{3\pi} (a^2\sin^2 t + a^2\cos^2 t)\cdot k\cdot\sqrt{a^2+b^2}\ dt = a^2k\sqrt{a^2+b^2}\int_0^{3\pi} dt$$

$$= a^2k\sqrt{a^2+b^2}\cdot 3\pi.$$

EXERCISES 18.3, page 907

1. We seek $f(x,y)$ such that $\nabla f = \underline{F}(x,y) = (3x^2y+2)\underline{i} + (x^3+4y^3)\underline{j}$. Since $\frac{\partial}{\partial y}(3x^2y+2) = 3x^2 = \frac{\partial}{\partial x}(x^3+4y^3)$, there might be such f. (Note: If $M_y \neq N_x$ then there can not be such f. Although at this stage of the game $M_y = N_x$ does not imply f exists, this equality does suggest the possible existence of f.) If such f exists, it must satisfy $f_x = 3x^2y + 2$ and $f_y = x^3+4y^3$. From the 1st, $f = x^3y + 2x + g(y)$ from which, $f_y = x^3 + g'(y)$. Thus to satisfy the 2nd equation, we must have $x^3 + g'(y) = x^3 + 4y^3$, or $g'(y) = 4y^3$. Thus $g(y) = y^4 + c$ and $f(x,y) = x^3y+2x+y^4+c$ is our desired function and the integral is independent of path.

4. Here $M = -y(x^2+y^2)^{-1/2}$, $N = x(x^2+y^2)^{-1/2}$. $M_y = x^2(x^2+y^2)^{-3/2} - (x^2+y^2)^{-1/2}$ $= (x^2+y^2)^{-3/2}(y^2-x^2-y^2) = -x^2(x^2+y^2)^{-3/2}$. In a similar manner we find $N_x = y^2(x^2+y^2)^{-3/2} \neq M_y$ and the integral is not independent of path.

7. If f is to satisfy $f_x = y\sec^2x - ze^x$, $f_y = \tan x$ and $f_z = -e^x$ then, integrating the 1st, $f = y\tan x = ze^x + g(y,z)$ and $f_y = \tan x + g_y$. Comparing with the f_y equation we see that $g_y = 0$ and $g(y,z) = h(z)$. Thus $f(x,y,z) = y\tan x - ze^x + h(z)$ and $f_z = -e^x + h'(z)$ from which we see $h'(z) = 0$ by comparing this with the f_z equation above. Thus $h(z) = c$ and $f(x,y,z) = y\tan x - z\,e^x + c$ is the desired potential function and the integral is independent of path.

10. If f is to satisfy $f_x = e^{-z}$, $f_y = 2y$, $f_z = xe^{-z}$ then from the 1st, $f = x\,e^{-z} + g(y,z)$ and $f_y = g_y$. Comparing the 2nd equation above, $g_y = 2y$ and $g(y,z) = y^2 + h(z)$. Thus $f = xe^{-z} + y^2 + h(z)$ and $f_z = -xe^{-z} + h'(z)$. In order that the 3rd equation, above, be satisfied, we must have $h'(z) = 2xe^{-z}$ which is absurd since $h(z)$ is a function of z alone, not x and z. Thus no potential function exists for $\underline{F}$ and the integral is not independent of path.

13. Setting $f_x = 6xy^3 + 2z^2$, $f_y = 9x^2y^2$, $f_z = 4xz+1$, then $f = 3x^2y^3 + 2xz^2 + g(y,z)$. Then $f_y = 9x^2y^2 + g_y = 9x^2y^2 \Rightarrow g_y = 0$ and $g(y,z) = h(z)$. Thus $f = 3x^2y^3 + 2xz^2 + h(z)$ and $f_z = 4xz + h'(z) = 4xz + 1 \Rightarrow h'(z) = 1 \Rightarrow h(z) = z+c$ so that $f(x,y,z) = 3x^2y^3 + 2xz^2 + z + c$ is a potential function.

$$\int_{(1,0,2)}^{(-2,1,3)} \nabla f\cdot d\underline{r} = f(-2,1,3)-f(1,0,2) = (-21+c)-(10+c) = -31.$$

16. To be directed away from the origin, $\underline{F}$ must be of the form $\underline{F}(x,y,z) = c(x,y,z)\underline{r}$ where $\underline{r} = x\underline{i} + y\underline{j} + z\underline{k}$ and $c(x,y,z) > 0$. Since $|\underline{F}(x,y,z)| = c(x,y,z)|\underline{r}| = c(x,y,z)\sqrt{x^2+y^2+z^2}$, this magnitude is proportional to the distance of (x,y,z) from the origin (this distance is $\sqrt{x^2+y^2+z^2}$) if c is a constant. Thus $\underline{F} = cx\underline{i} + cy\underline{j} + cz\underline{k}$ and $\underline{F}$ will be conservative if there exists a potential function $f(x,y,z)$ such that $\nabla f = \underline{F}$. Obviously $f(x,y,z) = \frac{c}{2}(x^2+y^2+z^2) = \frac{c}{2}\,\underline{r}\cdot\underline{r}$ fills the bill.

19. Here $M = 2xy$, $N = x^2+z^2$, $P = yz$ and $M_y = N_x = 2x$, $M_z = P_x = 0$, but $N_z = 2z$ whereas $P_y = z$. There is no region in three-dimensional space where $P_y = N_z$ and so the integral is not independent of path in general. (It would be independent of path on the xy-plane, $z = 0$.)

EXERCISES 18.4, page 916

1. $M = x^2 + y$, $N = xy^2 \Longrightarrow N_x = y^2$, $M_y = 1$, and $\oint_C (x^2+y^2)dx + xy^2dy = \iint_R (y^2-1)dA$, where R is the region enclosed by C, $R = \{(x,y)\colon -1 \le y \le 0,\ y^2 \le x \le -y\}$. Thus the value is

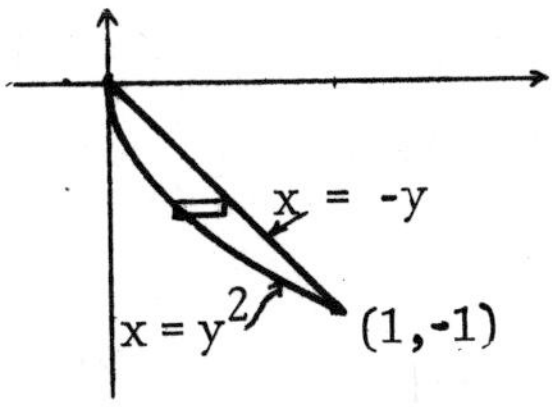

$$\int_{-1}^{0}\int_{y^2}^{-y} (y^2-1)\,dx\,dy = \int_{-1}^{0} (y^2-1)(-y-y^2)dy =$$

$$\int_{-1}^{0} (y+y^2-y^3-y^4)dy = 0-\left(\frac{1}{2} - \frac{1}{3} - \frac{1}{4} + \frac{1}{5}\right) = -\frac{7}{60}\,.$$

4. The edges of the triangle have equations $x = y$, $x = 4-y$, $y = 1$ from the given vertices. Thus the region R enclosed by C is of the form, $R = \{(x,y)\colon 1 \le y \le 2,\ y \le x \le 4-y\}$. Next $M = y^{1/2}$, $N = x^{1/2} \Longrightarrow N_x = \frac{1}{2}x^{-1/2}$, $M_y = \frac{1}{2}y^{-1/2}$ and $\oint_C \sqrt{y}\,dx + \sqrt{x}\,dy = \iint_R \left(\frac{1}{2}x^{-1/2} - \frac{1}{2}y^{-1/2}\right)dA = \frac{1}{2}\int_1^2\int_y^{4-y} (x^{-1/2} - y^{-1/2})\,dx\,dy =$

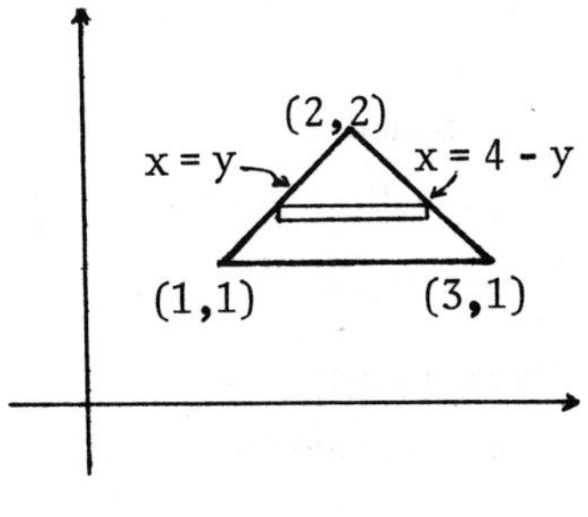

$$\frac{1}{2}\int_1^2 \left[2x^{1/2} - y^{-1/2}x\right]_y^{4-y} dy = \frac{1}{2}\int_1^2 \left[2(4-y)^{1/2} - y^{-1/2}(4-y) - 2y^{1/2} + y^{-1/2}y\right]dy =$$

$$\int_1^2 \left[(4-y)^{1/2} - 2y^{-1/2}\right]dy = -\frac{2}{3}(4-y)^{3/2} - 4y^{1/2}\Big]_1^2 = -\frac{2}{3}\cdot 2^{3/2} - 4\cdot 2^{1/2} + \frac{2}{3}\cdot 3^{3/2} + 4$$

$= \frac{-16\sqrt{2}}{3} + 2\sqrt{3} + 4.$

7. The equations for the edges of the triangle are $y = x$, $y = -2x+6$, and $y = -\frac{1}{2}x + \frac{3}{2}$ from the given vertices. The region R enclosed by C must be broken up into 2 subregions no matter which order of integration is selected. We choose $R = R_1 \cup R_2$ where $R_1 = \{(x,y): 1 \le x \le 2, -\frac{1}{2}x + \frac{3}{2} \le y \le x\}$, the left portion, and $R_2 = \{(x,y): 2 \le x \le 3, -\frac{1}{2}x + \frac{3}{2} \le y \le -2x+6\}$, the right portion. From $M = xy$, $N = \sin y$ we get $M_y = x$, $N_x = 0$. Thus

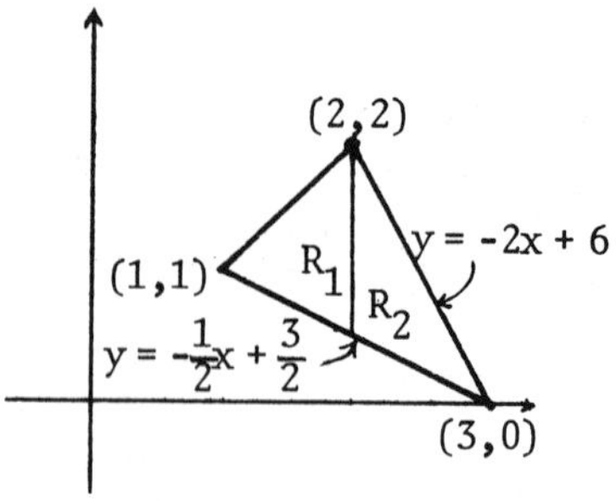

$$\oint_C xy\,dx + \sin y\,dy = \iint_R x\,dA = \iint_{R_1} x\,dA + \iint_{R_2} x\,dA = \int_1^2 \int_{-\frac{1}{2}x+\frac{3}{2}}^{x} x\,dy\,dx +$$

$$\int_2^3 \int_{-\frac{1}{2}x+\frac{3}{2}}^{-2x+6} x\,dy\,dx = \int_1^2 x(x + \frac{1}{2}x - \frac{3}{2})\,dx + \int_2^3 x(-2x+6+\frac{1}{2}x - \frac{3}{2})\,dx =$$

$$\int_1^2 (\frac{3}{2}x^2 - \frac{3}{2}x)\,dx + \int_2^3 (-\frac{3}{2}x^2 + \frac{9}{2}x)\,dx = \frac{1}{2}x^3 - \frac{3}{4}x^2\Big]_1^2 + [-\frac{1}{2}x^3 + \frac{9}{4}x^2]_2^3 =$$

$$[1 - (-\frac{1}{4})] + [\frac{27}{4} - 5] = \frac{5}{4} + \frac{7}{4} = 3.$$

10. $M = x^2 + y^2$, $N = 2xy \Longrightarrow M_y = 2y$, $N_x = 2y$ and since $N_x - M_y = 0$, the value of the integral is $\iint_R (N_x - M_y)\,dA = \iint_R 0\,dA = 0$.

13. The boundary, C, of the region will be expressed as $C = C_1 \cup C_2$ where C_1 is the parabola $y = 4x^2$ from (0,0) to (4,64) and C_2 is the line $y = 16x$ from (4,64) back to (0,0). We must choose these directions on C_1 and C_2 so that the positive direction around C results. Then on C_1 $dy = 8x\,dx$ and x varies from 0 to 4; on C_2 $dy = 16\,dx$ and x varies from 4 to 0. Then

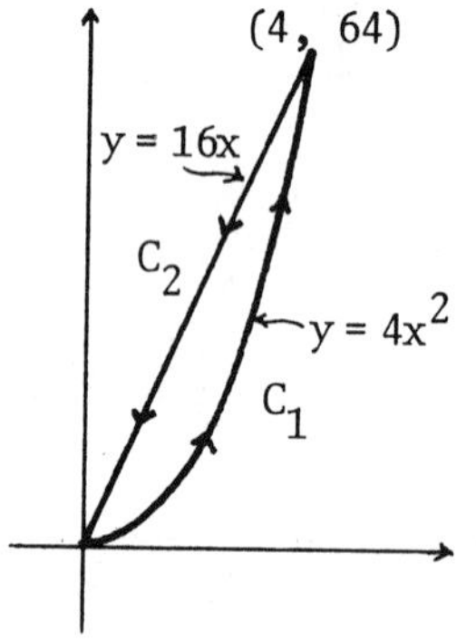

$$A = \frac{1}{2}\oint_C x\,dy - y\,dx = \frac{1}{2}\int_{C_1} x\,dy - y\,dx +$$

$$\frac{1}{2}\int_{C_2} x\,dy - y\,dx = \frac{1}{2}\int_0^4 [x(8x)-4x^2]\,dx + \frac{1}{2}\int_4^0 [x(16)-16x]\,dx = \frac{1}{2}\int_0^4 4x^2\,dx =$$

$$\frac{2}{3}x^3\Big]_0^4 = \frac{128}{3}.$$

16. If $\underline{F} = M\underline{i} + N\underline{j}$, then independence of path $\Longrightarrow M_y = N_x \Longrightarrow \oint_C \underline{F}\cdot d\underline{r} = \oint_C M\,dx +$

$N\,dy = \iint_R (N_x - M_y)\,dA = \iint_R 0\,dA = 0.$

19. Placing the origin at the center of the circle and using the upper half, the boundary C will be expressed as $C = C_1 \cup C_2$ where C_1 is the circular path from (a,0) to (-a,0) given by x = a cos t, y = a sin t, for t varying from 0 to π, and C_2 is the path on the x axis from (-a,0) to (a,0) given by x = t, y = 0 for t varying from -a to a. Using $A = \frac{1}{2}\pi a^2$ and #17 we obtain:

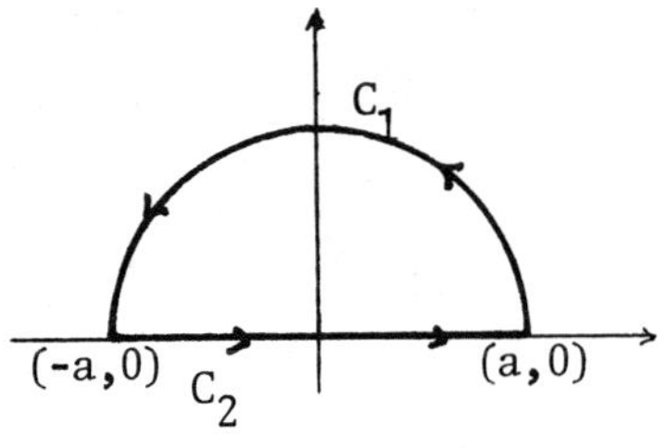

$$\overline{x} = \frac{1}{\pi a^2}\oint_C x^2\,dy = \frac{1}{\pi a^2}\left(\int_{C_1} x^2\,dy + \int_{C_2} x^2\,dy\right) = \frac{1}{\pi a^2}\left(\int_0^{\pi} (a^2\cos^2 t)(a\cos t)\,dt +\right.$$

$$\left.\int_{-a}^{a} t^2\cdot 0\,dt\right) = \frac{1}{\pi a^2}\cdot a^3\int_0^{\pi}\cos^3 t\,dt = \frac{a}{\pi}\left[\sin t - \frac{\sin^3 t}{3}\right]_0^{\pi} = 0.$$

$$\overline{y} = -\frac{1}{\pi a^2}\oint_C y^2\,dx = -\frac{1}{\pi a^2}\left(\int_{C_1} y^2\,dx + \int_{C_2} y^2\,dx\right) =$$

$$-\frac{1}{\pi a^2}\left(\int_0^{\pi} a^2\sin^2 t(-a\sin t)\,dt + \int_{-a}^{a} 0\,dt\right) = \frac{a}{\pi}\int_0^{\pi}\sin^3 t = \frac{a}{\pi}\left[-\cos t + \frac{\cos^3 t}{3}\right]_0^{\pi}$$

$$= \frac{a}{\pi}\cdot\frac{4}{3}.$$

EXERCISES 18.5, page

1.
$$\nabla \times \underline{F} = \begin{vmatrix} \underline{i} & \underline{j} & \underline{k} \\ \partial/\partial x & \partial/\partial y & \partial/\partial z \\ x^2 z & y^2 z & y+2z \end{vmatrix} = (1-0)\underline{i} - (0-x^2)\underline{j} + (y^2-0)\underline{k}$$

$$\nabla\ \underline{F} = \frac{\partial}{\partial x}(x^2 z) + \frac{\partial}{\partial y}(y^2 x) + \frac{\partial}{\partial z}(y+2z) = 2xz + 2xy + 2.$$

2. $$\nabla \times \underline{F} = \begin{vmatrix} \underline{i} & \underline{j} & \underline{k} \\ \partial/\partial x & \partial/\partial y & \partial/\partial z \\ 3x+y & xy^2z & xz^2 \end{vmatrix} = (0-xy^2)\underline{i} - (z^2-0)\underline{j} + (y^2z-1)\underline{k}$$

$$\nabla \cdot \underline{F} = \frac{\partial}{\partial x}(3x+y) + \frac{\partial}{\partial y}(xy^2z) + \frac{\partial}{\partial z}(xz^2) = 3 + 2xyz + 2xz.$$

7. With $\underline{F} = M\underline{i} + N\underline{j} + P\underline{k}$, $\nabla\times(f\underline{F}) =$

$$\begin{vmatrix} \underline{i} & \underline{j} & \underline{k} \\ \partial/\partial x & \partial/\partial y & \partial/\partial z \\ fM & fN & fP \end{vmatrix} = [\frac{\partial}{\partial y}(fP) - \frac{\partial}{\partial z}(fN)]\underline{i} - [\frac{\partial}{\partial x}(fP) - \frac{\partial}{\partial z}(fM)]\underline{j} + [\frac{\partial}{\partial x}(fN) - \frac{\partial}{\partial y}(fM)]\underline{k}.$$

Now, using the product rule and rearranging, the result follows.

10. Recall that the mixed 2nd partial derivatives are equal if continuous, which is the case here. Thus div curl $\underline{F}$ = div$[(P_y-N_z)\underline{i} - (P_x-M_z)\underline{j} + (N_x-M_y)\underline{k}]$ = $(P_{yz}-N_{zx})-(P_{xy}-M_{zy}) + (N_{xz}-M_{yz})$. The mixed derivatives for each function occur with opposite signs, and, being equal, they all cancel resulting in 0.

13. The 1st 2 conclusions are very easy to get. For the 3rd, $|\underline{r}| = (x^2+y^2+z^2)^{1/2}$ $\Rightarrow \frac{\partial|\underline{r}|}{\partial x} = \frac{x}{(x^2+y^2+z^2)^{1/2}} = \frac{x}{|\underline{r}|}$ and similarly $\frac{\partial}{\partial y}|\underline{r}| = \frac{y}{|\underline{r}|}$, $\frac{\partial}{\partial z}|\underline{r}| = \frac{z}{|\underline{r}|}$.

Thus $\nabla|\underline{r}| = \frac{\partial}{\partial x}|\underline{r}|\underline{i} + \frac{\partial}{\partial y}|\underline{r}|\underline{j} + \frac{\partial}{\partial z}|\underline{r}|\underline{k} = \frac{x}{|\underline{r}|}\underline{i} + \frac{y}{|\underline{r}|}\underline{j} + \frac{z}{|\underline{r}|}\underline{k} = \frac{1}{|\underline{r}|}(x\underline{i} + y\underline{j} + z\underline{k}) = \frac{\underline{r}}{|\underline{r}|}$.

16. By #7 above, $\nabla\times\underline{F} = \nabla\times(\frac{c}{r^k})\underline{r} = \frac{c}{r^k}(\nabla\times\underline{r}) + (\nabla\frac{c}{r^k})\times\underline{r}$. The 1st term is $\underline{0}$ by #13 above. To evaluate the 2nd term we compute, using the chain rule, $\frac{\partial}{\partial x}(\frac{c}{r^k}) = \frac{-kcr_x}{r^{k+1}} = \frac{-kcx}{r^{k+2}}$ (See #13 above). Similarly $\frac{\partial}{\partial y}(\frac{c}{r^k}) = \frac{-kcy}{r^{k+2}}$, $\frac{\partial}{\partial z}(\frac{c}{r^k}) = \frac{-kcz}{r^{k+2}}$ so that $\nabla(\frac{c}{r^k}) = \frac{-kc}{r^{k+2}}(x\underline{i} + y\underline{j} + z\underline{k}) = \frac{-kc}{r^{k+2}}\underline{r}$. Thus the 2nd term in the initial calculation is also $\underline{0}$ since $\underline{r}\times\underline{r} = \underline{0}$, and the entire result is $\underline{0}$ as desired.

19. With $f(x,y,z) = (x^2+y^2+z^2)^{-1/2} = \frac{1}{r}$, $f_x = -\frac{x}{r^3}$, $f_y = -\frac{y}{r^3}$, $f_z = -\frac{z}{r^3}$. (See #16 above with c = k = 1.) Thus

$$f_{xx} = \frac{-r^3 + x\frac{\partial}{\partial x}r^3}{r^6} = \frac{-r^3 + x(3r^2r_x)}{r^6} = \frac{-r^3 + 3rx}{r^6} = \frac{3x^2 - r^2}{r^5}.$$

(We used $r_x = x/r$.) Similar formulas hold for f_{yy} and f_{zz} with y and z replacing the x. Finally,

$$\nabla^2 f = \frac{3(x^2+y^2+z^2) - 3r^2}{r^5} = 0.$$

EXERCISES 18.6, page 927

1. S is given by $z = \sqrt{a^2-x^2-y^2} = f(x,y)$ and

$$f_x = \frac{-x}{\sqrt{a^2-x^2-y^2}},\ f_y = \frac{-y}{\sqrt{a^2-x^2-y^2}},$$

$$\sqrt{1+f_x^2+f_y^2} = \sqrt{1 + \frac{x^2}{a^2-x^2-y^2} + \frac{y^2}{a^2-x^2-y^2}} =$$

$$\sqrt{\frac{a^2-x^2-y^2+x^2+y^2}{a^2-x^2-y^2}} = \frac{a}{\sqrt{a^2-x^2-y^2}}.$$ The region R in the xy-plane below S is the disc $x^2+y^2 \le a^2$; so polar coordinates are indicated. In R, $0 \le r \le a$, $0 \le \theta \le 2\pi$.

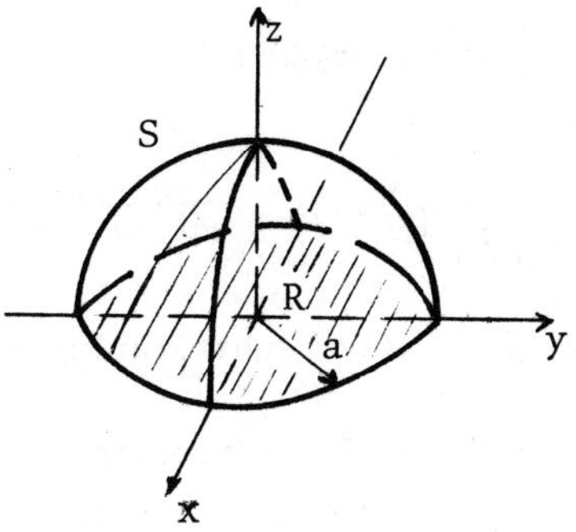

$$I = \iint_S f(x,y,z)\, dS = \iint_S x^2\, dS =$$

$$\iint_R x^2\sqrt{1+f_x^2+f_y^2}\, dA = \int_0^{2\pi}\int_0^a r^2\cos^2\theta \cdot \frac{a}{\sqrt{a^2-r^2}} \cdot r\, dr\, d\theta.$$ To find

$\displaystyle\int_0^a \frac{r^3}{\sqrt{a^2-r^2}}\, dr$ let $r = a \sin t$ to obtain $\displaystyle\int_0^{\pi/2} a^3 \sin^3 t\, dt = \frac{2}{3} a^3$ and

$$I = \frac{2}{3} a^4 \int_0^{2\pi} \cos^2\theta\, d\theta = \frac{2}{3} a^4 \cdot \pi.$$

4. S is given by $z = \frac{x^2}{2} + \frac{y^2}{2} = f(x,y)$ so that $\sqrt{1+f_x^2+f_y^2} = \sqrt{1+x^2+y^2}$. The region R is inside the circle $x^2+y^2 = 2y$ or $r = 2\sin\theta$ in polar form, i.e. $R = \{(r,\theta): 0 \le \theta \le \pi,\ 0 \le r \le 2\sin\theta\}$.

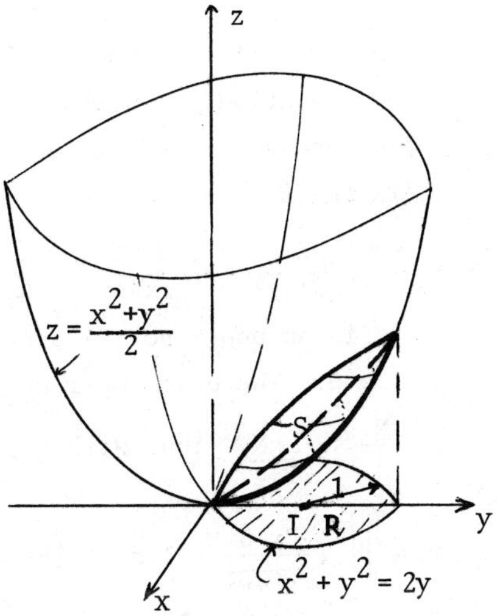

$$I = \iint_S f(x,y,z)\, dS = \iint_R \sqrt{x^2+y^2+1}\,\sqrt{x^2+y^2+1}\, dA$$

$$= \int_0^{\pi}\int_0^{2\sin\theta} (r^2+1)\, r\, dr\, d\theta =$$

$$\int_0^{\pi} \left[\frac{r^4}{4} + \frac{r^2}{2}\right]_0^{2\sin\theta} d\theta = \int_0^{\pi} (4\sin^4\theta +$$

$$2\sin^2\theta)\, d\theta = \int_0^{\pi} [(1-\cos 2\theta)^2 + (1-\cos 2\theta)]\, d\theta$$

$$= \int_0^{\pi} \left(\frac{5}{2} - 3\cos 2\theta + \frac{1}{4}\cos 2\theta\right) d\theta = \frac{5\pi}{2}.$$

7. (a) Rewrite the equation of S in the form $x = 2-y/4 = k(y,z)$. Then $k_y = -1/4$ and $k_z = 0$. The projection of S onto the yz-plane is the rectangular region $R_a = \{(y,z): 0 \le y \le 8,\ 0 \le z \le 6\}$. Thus

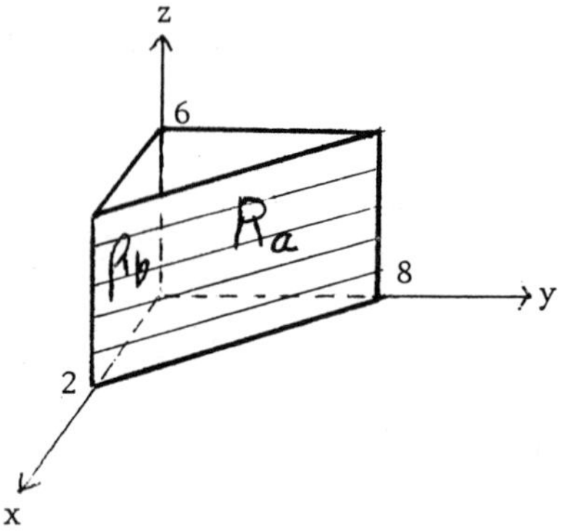

$$\iint_S (x^2-2y+z)\,dS =$$

$$\int_0^8 \int_0^6 (x^2-2y+z)\sqrt{(-1/4)^2 + 0^2 + 1}\ dz\ dy,$$

which, after substituting $x = 2-y/4$, becomes $\int_0^8 \int_0^6 [(2-y/4)^2-2y+z]\sqrt{17/16}\ dz\ dy$,

which easily reduces to the answer given in the text.

(b) Now rewrite the S equation in the form $y = 8-4x = k(x,z)$. Then $k_x = -4$ and $k_z = 0$. The projection of S onto the xz-plane is the rectangular region $R_b = \{(x,z): 0 \le x \le 2,\ 0 \le z \le 6\}$. Replacing y by $8-4x$ (from the S equation above) in the integrand, we obtain

$$\iint_S (x^2-2y+z)\,dS = \int_0^2 \int_0^6 [x^2-2(8-4x)+z]\sqrt{(-4)^2 + 0^2 + 1}\ dz\ dx,$$

which reduces to the answer given in the text.

13. With S given by $z = \sqrt{x^2+y^2} = f(x,y)$, we have $f_x = \dfrac{x}{\sqrt{x^2+y^2}}$, $f_y = \dfrac{y}{\sqrt{x^2+y^2}}$,

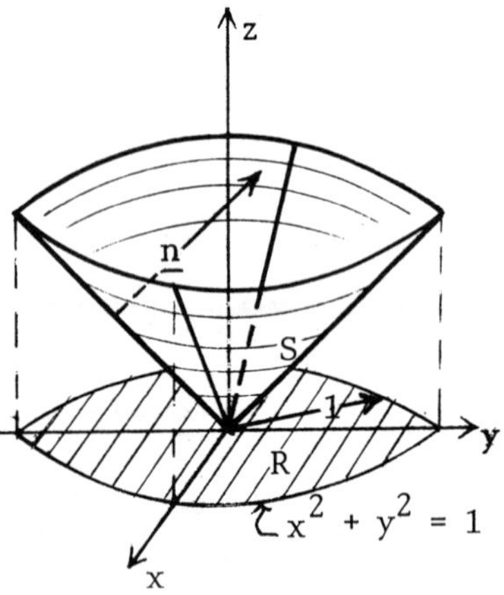

$$\sqrt{f_x^{\,2} + f_y^{\,2} + 1} = \sqrt{\frac{x^2}{x^2+y^2} + \frac{y^2}{x^2+y^2} + 1} = \sqrt{2}.$$

Now, recall from Sec. 16.7 that if a surface has equation $g(x,y,z) = 0$, then ∇g is normal to the surface. Writing the equation for S in the form $z - \sqrt{x^2+y^2} = 0$ we find that

$\underline{N} = \dfrac{-x}{\sqrt{x^2+y^2}}\underline{i} - \dfrac{y}{\sqrt{x^2+y^2}}\underline{j} + \underline{k}$ is normal to S.

It is an upper normal since $\underline{k}$ has coefficient $1 > 0$. The unit upper normal $\underline{n} = \dfrac{\underline{N}}{|\underline{N}|} = \dfrac{1}{\sqrt{2}}\underline{N}$. Thus with $\underline{F} = 2\underline{i} + 5\underline{j} + 3\underline{k}$, $I = \iint_S \underline{F}\cdot\underline{n}\,dS = \iint_S \dfrac{1}{\sqrt{2}}\left(\dfrac{-2x-5y}{\sqrt{x^2+y^2}} + 3\right) dS$

$= \iint_R \left(\dfrac{-2x-5y}{\sqrt{x^2+y^2}} + 3\right)dA$. (From the 1st line, $\sqrt{f_x^{\,2} + f_y^{\,2} + 1} = \sqrt{2}$.) R is the

circular region $x^2+y^2 \le 1$ or $\{(r,\theta): 0 \le r \le 1,\ 0 \le \theta \le 2\pi\}$, so that

$$I = \int_0^1 \int_0^{2\pi} \left(\frac{-2r\cos\theta - 3r\sin\theta}{r} + 3\right) r\, d\theta\, dr = 6\pi \int_0^1 r\, dr = 3\pi.$$

(The $\cos\theta$ and $\sin\theta$ terms yield 0.)

16. By the remarks after (18.27), if $g(x,y,z)$ is the area density at (x,y,z), then $M = \iint_S g(x,y,z)\, dS$. Here, $g(x,y,z) = z^2$ and, from Example 1, we have $z^2 = x^2+y^2$, $dz = \sqrt{2}\, dx\, dy$ and $R = \{(r,\theta): 1 \le r \le 4,\ 0 \le \theta \le 2\pi\}$ in polar form. Thus $M = \iint_S z^2 dS = \iint_R (x^2+y^2)\sqrt{2}\, dx\, dy = \sqrt{2}\int_1^4 \int_0^{2\pi} r^2 \cdot r\, d\theta\, dr =$

$$2\sqrt{2}\pi \int_1^4 r^3\, dr = \frac{255\sqrt{2}\pi}{2}.$$

By symmetry, the center of gravity will be on the z-axis, and $\bar{x} = \bar{y} = 0$. To obtain $\bar{z}$, we first derive a formula for M_{xy} as follows. Let the surface, S, be partitioned as described in Section 17.9. The ith piece has area $\Delta S_i \approx \Delta T_i$. (See page 878 and Figure 17.42.) If (x_i,y_i,z_i) is a point of tangency on the ith piece, the mass of this small piece of S is $\approx g(x_i,y_i,z_i)\Delta T_i$ and its moment with respect to the xy-plane is $\approx z_i g(x_i,y_i,z_i)\Delta T_i$. Summing and passing to the limit, we obtain $M_{xy} = \iint_S zg(x,y,z)dS$ and $\bar{z} = M_{xy}/M$. Using $g(x,y,z) = z^2$ and $z = (x^2+y^2)^{1/2}$, we obtain $M_{xy} = \iint_S z\cdot z^2 dS =$

$$= \sqrt{2}\iint_R (x^2+y^2)^{3/2}\, dx\, dy = \sqrt{2}\int_1^4 \int_0^{2\pi} r^3\cdot r\, d\theta\, dr = 2\sqrt{2}\pi\int_1^4 r^4 dr = \frac{2046\sqrt{2}\pi}{5},$$

and $\bar{z} = \dfrac{2046\sqrt{2}\pi}{5} \cdot \dfrac{2}{255\sqrt{2}\pi} = \dfrac{1364}{425} \approx 3.21.$

For the moment of inertia with respect to the z-axis, we partition as above. The moment of inertia of the ith piece about the z-axis is $\approx (x_i^2+y_i^2)g(x_i,y_i,z_i)\Delta T_i$. Summing and passing the limit, we obtain $I_z = \iint_S (x^2+y^2)g(x,y,z)dS$. Here, using $g(x,y,z) = z^2$ and $z^2 = x^2+y^2$ on S, we get $I_z = \iint_S (x^2+y^2)z^2\, dS = \sqrt{2}\iint_R (x^2+y^2)^2\, dx\, dy = \sqrt{2}\int_1^4 \int_0^{2\pi} r^4\cdot r\, d\theta\, dr =$

$$2\sqrt{2}\pi\int_1^4 r^5 dr = 1365\sqrt{2}\pi.$$

EXERCISES 18.7, page 936

1. div $\underline{F}$ = y cos x + 2yz + 3 and $\iint_S \underline{F}\cdot\underline{n}\,dS = \iiint_Q \text{div}\,\underline{F}\,dV$ where Q is the cube bounded by the given planes. Thus $\iint_S \underline{F}\cdot\underline{n}\,dS = \int_{-1}^{1}\int_{-1}^{1}\int_{-1}^{1} (y\cos x + 2yz + 3)\,dzdydx = \int_{-1}^{1}\int_{-1}^{1} (2y\cos x + 6)\,dydz = \int_{-1}^{1} 12\,dx = 24.$

4. Q is the region between the paraboloid $z = x^2+y^2$ and the plane z = 9. In Q, z varies from x^2+y^2 to 9, and the region R in the xy-plane is enclosed by the circle $x^2+y^2 = 9$. In cylindrical coordinates Q = $\{(r,\theta,z): 0 \le r \le 3,\ 0 \le \theta \le 2\pi,\ r^2 \le z \le 9\}$. Since div $\underline{F}$ = 2y + 2z we obtain $\iint_S \underline{F}\cdot\underline{n}\,dS = \iiint_Q \text{div}\,\underline{F}\,dV =$

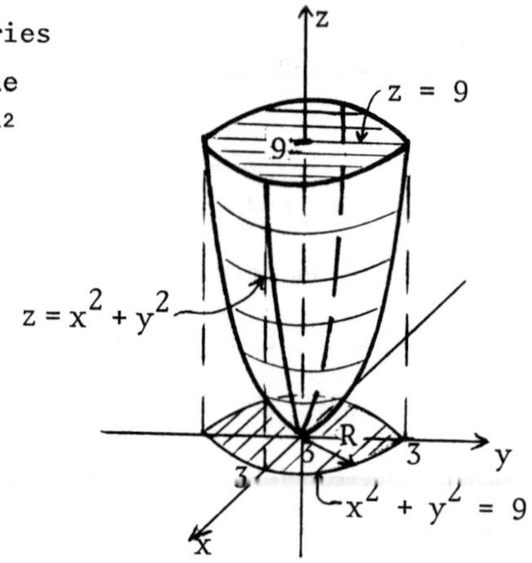

$\int_0^3\int_0^{2\pi}\int_{r^2}^{9} (2r\sin\theta + 2z)r\,dz\,d\theta\,dr =$

$\int_0^3\int_0^{2\pi} [2r^2\sin\theta(9-r^2) + r(81-r^4)]d\theta\,dr =$

$2\pi\int_0^3 (81r - r^5)dr = 2\pi[\frac{81}{2}r^2 - \frac{r^6}{6}]_0^3 = 486\pi.$

7. First, let's compute $\iiint_Q \text{div}\,\underline{F}\,dV$. Here div $\underline{F}$ = 2, and this integral is just twice the volume of Q. Q is a right circular cylinder of base area $\pi\cdot 1^2 = \pi$ and "height" 2 (from the yz-plane, x = 0, to x = 2). Its volume is 2π and the integral is 4π. Now, the closed surface S consists of 3 pieces: S_1, the back face in the yz-plane given by $y^2+z^2 \le 1$, x = 0 for which the outer unit normal is $-\underline{i}$; S_2, the curved cylindrical surface, $y^2+z^2 = 1$, $0 \le x \le 2$

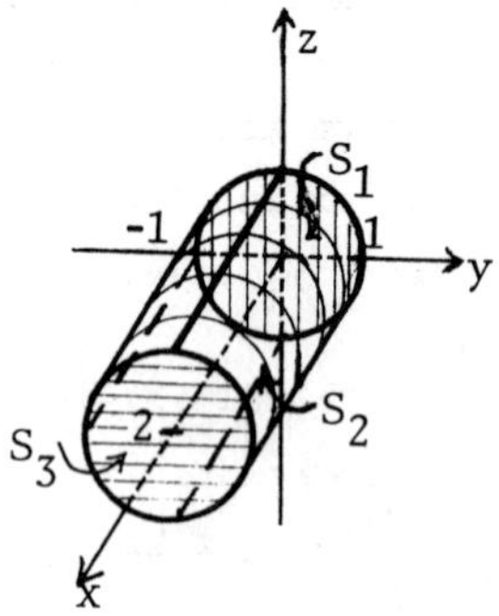

for which the outer unit normal is $\frac{\nabla g}{|\nabla g|} = \frac{<0,2y,2z>}{\sqrt{4y^2+4z^2}} = <0,y,z>$ (remember y^2 +

$z^2 = 1$ on S_2); and S_3, the front face $y^2+z^2 \le 1$, $x = 2$ for which the outer unit normal is $\mathbf{i}$.

Since $\iint_S = \iint_{S_1} + \iint_{S_2} + \iint_{S_3}$, we'll calculate each one separately and then add.

$\iint_{S_1} \underline{F}\cdot\underline{n}\, dS = \iint_{S_1} \underline{F}\cdot(-\underline{i})\, dS = \iint_{S_1} -(x+z)\, dS = -\iint_{S_1} z\, dS$ since $x = 0$ on S_1. Now, S_1 is itself the region R in the yz-plane given by $y^2+z^2 \le 1$ or in polar coordinates ($y = r\cos\theta$, $z = r\sin\theta$), $S_1 = R = \{(r,\theta): 0 \le r \le 1, 0 \le \theta \le 2\pi\}$. Thus $\iint_{S_1} \underline{F}\cdot\underline{n}\, dS = -\int_0^1 \int_0^{2\pi} (r\sin\theta)\, r\, d\theta\, dr = 0$.

Next $\iint_{S_2} \underline{F}\cdot\underline{n}\, dS = \iint_{S_2} [(x+z)0 + (y+z)y + (x+y)z]\, dS = \iint_{S_2} [y^2 + 2yz + xz]\, dS$.

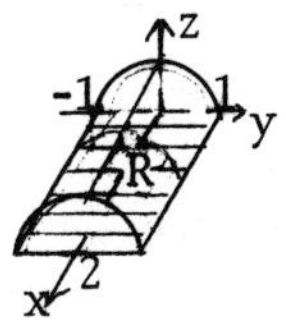

The region R in the xy-plane above and below which S_2 extends is the rectangle $0 \le x \le 2$, $-1 \le y \le 1$. The top half of S_2 has equation $z = \sqrt{1-y^2}$ so that $f_x = 0$, $f_y = \dfrac{-y}{\sqrt{1-y^2}}$ and $\sqrt{f_x^{\,2} + f_y^{\,2} + 1} = \sqrt{\dfrac{y^2}{1-y^2} + 1} = \dfrac{1}{\sqrt{1-y^2}}$.

The bottom half of S_2 has equation $z = -\sqrt{1-y^2}$ so that $f_x = 0$, $f_y = \dfrac{y}{\sqrt{1-y^2}}$ and $\sqrt{f_x^{\,2} + f_y^{\,2} + 1} = \dfrac{1}{\sqrt{1-y^2}}$ also.

Now $\iint_{S_2} = \iint_{\text{top}} + \iint_{\text{bottom}} = \int_{-1}^1 \int_0^2 (y^2 + 2y\sqrt{1-y^2} + x\sqrt{1-y^2})\dfrac{1}{\sqrt{1-y^2}}\, dx\, dy + \int_{-1}^1 \int_0^2 (y^2 - 2y\sqrt{1-y^2} - x\sqrt{1-y^2})\dfrac{1}{\sqrt{1-y^2}}\, dx\, dy$. (We have, of course, substituted $z = \sqrt{1-y^2}$ in the 1st, $z = -\sqrt{1-y^2}$ in the 2nd.) The 2nd and 3rd terms cancel when the sum is taken, and $\iint_{S_2} = 2\int_{-1}^1 \int_0^2 \dfrac{y^2}{\sqrt{1-y^2}}\, dx\, dy = 4\int_{-1}^1 \dfrac{y^2}{\sqrt{1-y^2}}\, dy = 4\int_{-\pi/2}^{\pi/2} \sin^2\theta\, d\theta = 2\int_{-\pi/2}^{\pi/2} (1 - \cos 2\theta)\, d\theta = 2\pi$. (We let $y = \sin\theta$ in the 2nd last line.)

Finally, $\iint_{S_3} \underline{F}\cdot\underline{n}\, dS = \iint_{S_3} (x+z)\, dS$. The integral of z works out to be 0 as in the S_1 integral. On S_3, $x = 2$ and $\iint_{S_3} 2\, dS$ = twice the area of the circular region R in the yz-plane = 2π. The sum of these 3 surface integrals is $0 + 2\pi + 2\pi = 4\pi$, the same as the volume integral of div $\underline{F}$.

10. Hint: $D_{\underline{n}}f = \nabla f \cdot \underline{n}$ where $\underline{n}$ is a unit normal. Now use the divergence theorem and div $\nabla f = \nabla^2 f$.

13. Hint: By the divergence theorem $\iint_S f\underline{F}\cdot\underline{n}\, dS = \iiint_Q \text{div } f\underline{F}\, dV$. Compute div $f\underline{F}$, use $\nabla f = \underline{F}$ and div $\underline{F} = 0$.

16. Hint: div $\underline{a} = 0$ if $\underline{a}$ is constant.

EXERCISES 18.8, page 944

1. First we'll compute $\iint_S \text{curl } \underline{F}\cdot\underline{n}\, dS$. In the usual way, curl $\underline{F} = -2z\underline{i} - 2x\underline{j} - 2y\underline{k}$. S is the plane $x+y+z = 1$ and a normal to it is $\underline{i} + \underline{j} + \underline{k}$. A unit normal, pointing into the 1st octant, is $\frac{1}{\sqrt{3}}(\underline{i} + \underline{j} + \underline{k})$ and curl $\underline{F}\cdot\underline{n} = -\frac{2}{\sqrt{3}}(z + x + y)$. On S, this is just $-\frac{2}{\sqrt{3}}$ since $x+y+z = 1$ there. Writing this equation as $z = 1-x-y$ we find that

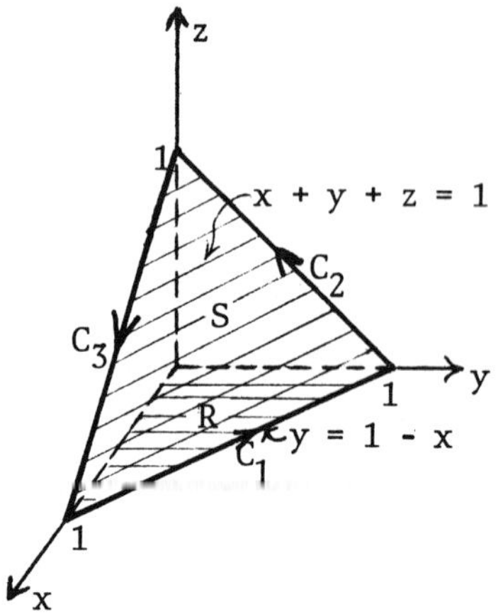

$\sqrt{f_x^2 + f_y^2 + 1} = \sqrt{(-1)^2 + (-1)^2 + 1} = \sqrt{3}$. The region R under S is $\{(x,y): 0 \le x \le 1,\ 0 \le y \le 1-x\}$ triangular in shape. Thus $\iint_S \text{curl } \underline{F}\cdot\underline{n}\, dS =$

$$\iint_S \left(-\frac{2}{\sqrt{3}}\right) dS = \iint_R \left(-\frac{2}{\sqrt{3}}\right)\sqrt{3}\, dA = -2\iint_R dA = -2\cdot(\text{Area of R}) = -2\cdot\frac{1}{2} = -1.$$

To calculate $\oint_C \underline{F}\cdot d\underline{r} = \oint_C y^2dx + z^2dy + x^2dz$, we express $C = C_1 \cup C_2 \cup C_3$, as shown in the sketch, to get the positive direction. C_1 is given by $x = 1-t$, $y = t$, $z = 0$, $0 \le t \le 1$ and $\int_{C_1} \underline{F}\cdot d\underline{r} = \int_0^1 [t^2(-1) + 0^2 + (1-t)^2\cdot 0]dt =$

$-\int_0^1 t^2\, dt = -\frac{1}{3}$. C_2 is given by $x = 0$, $y = 1-t$, $z = t$, $0 \le t \le 1$, and

$\int_{C_2} \underline{F}\cdot d\underline{r} = \int_0^1 [(1-t)^2\cdot 0 + t^2(-1) + 0^2]dt = -\int_0^1 t^2dt = -\frac{1}{3}$. C_3 is given by

$x = t,\ y = 0,\ z = 1-t,\ 0 \le t \le 1$ and $\int_{C_3} \underline{F}\cdot d\underline{r} = \int_0^1 [0^2 + (1-t)^2\cdot 0 + t^2(-1)]dt$

$= -\int_0^1 t^2 dt = -\frac{1}{3}$. Their sum is -1 which agrees with the surface integral.

4. curl $\underline{F} = \underline{0}$ and so $\iint_S \text{curl } \underline{F}\cdot\underline{n}\, dS = 0$. C can be parametrized by $x = \sin t$, $y = \cos t$, $z = 1$, $0 \le t \le 2\pi$ so that $\oint_C \underline{F}\cdot d\underline{r} = \oint_C x^2dx + y^2dy + z^2dz = \int_0^{2\pi} [\sin^2 t \cos t + \cos^2 t(-\sin t) = \frac{1}{3}[\sin^3 t + \cos^3 t]_0^{2\pi} = 0$.

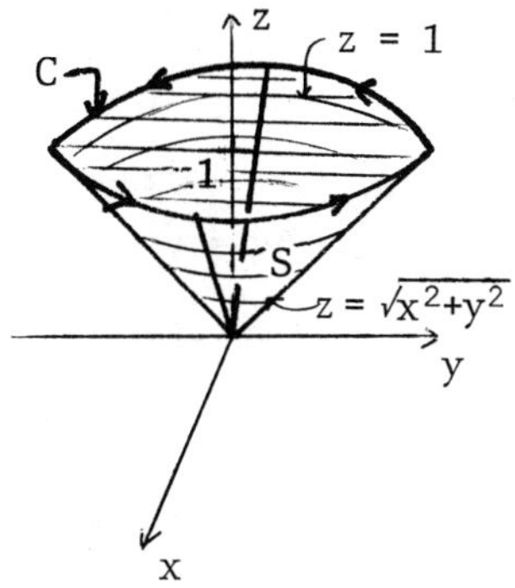

6. The value of the integral is -3/2.

7. C is the circle $x^2+y^2 = 4$ in the xy-plane. To obtain the positive direction along C, we use the parametric equations: $x = 2\cos t$, $y = 2\sin t$, $z = 0$, $0 \le t \le 2\pi$. Then $\iint_S \text{curl } \underline{F}\cdot\underline{n}\, dS = \oint_C \underline{F}\cdot d\underline{r} = \oint_C 2y\, dx + e^z dy - \arctan x\, dz =$

$\int_0^{2\pi} [2(2\sin t)(-2\sin t) + e^0(2\cos t) -$

$-\arctan(2\cos t)(0)]dt =$

$\int_0^{2\pi} (-8\sin^2 t + 2\cos t)dt = \int_0^{2\pi} (-4 + 4\cos 2t + 2\cos t)dt = -8\pi$.

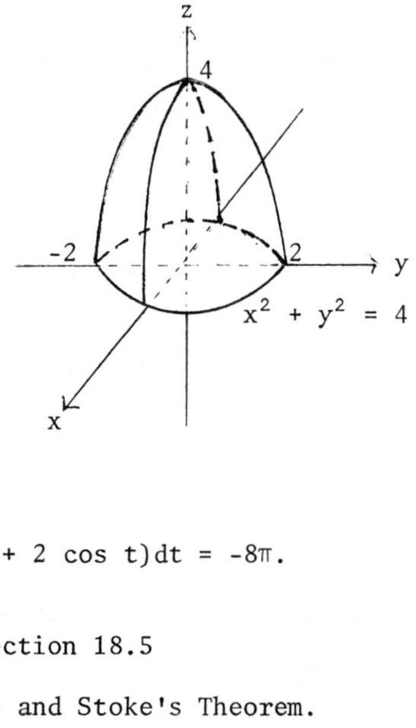

9. Hint: Use Stoke's Theorem, then #1 and #9 of Section 18.5

10. Hint: Let $\underline{F} = \underline{a} \times \underline{r}$ and use #14 of Section 18.5 and Stoke's Theorem.

13. $v_y = u_x \Rightarrow v_{yy} = u_{xy}$. $v_x = -u_y \Rightarrow v_{xx} = -u_{yx}$ and $\nabla^2 v = v_{xx} + v_{yy} = -u_{yx} + u_{xy} = 0$ since these mixed derivatives are equal if continuous.

EXERCISES 18.9, page 949

1. Recalling that the u-curves and v-curves are graphs in the xy-plane of the equations $u = f(x,y) = c$ and $v = g(x,y) = d$, respectively, the u-curves here

are given by $3x = c$, or $x = c/3$, vertical lines, and the v-curves by $5y = d$, or $y = d/5$, horizontal lines.

(b) To find the functions $\phi(u,v)$ and $\psi(u,v)$ which occur in T^{-1}, we must solve, if possible, the T equations, $u = f(x,y)$ and $v = g(x,y)$, for x and y. Here, $u = 3x$, $v = 5y \Rightarrow x = u/3 = \phi(u,v)$, $y = v/5 = \psi(u,v)$.

4. (a) The u-curves are graphs of $u = 5x-4y = c$, lines of slope 5/4, and the v-curves are graphs of $v = 2x+3y = d$, lines of slope -2/3.

(b) $$\left.\begin{array}{r} 5x-4y = u \\ 2x+3y = v \end{array}\right\} \Longrightarrow \left.\begin{array}{r} 15x-12y = 3u \\ 8x+12y = 4v \end{array}\right\}$$ Add to get $23x = 3u+4v$ so that $x = \frac{3}{23}u + \frac{4}{23}v = \phi(u,v)$. Using this in either of the original equations yields $y = -\frac{2}{23}u + \frac{5}{23}v = \psi(u,v)$.

7. (b) T is not one-to-one since different points in the xy-plane may have the same image in the uv-plane. For example, the points $(x,y) = (\pm 2, \pm 3)$ have the same image $(u,v) = (40,7)$.

10. (a) The u-curves are graphs of $u = \frac{y}{x^2+y^2} = c$. If $c \neq 0$ we obtain $x^2+y^2 = \frac{y}{c}$, a circle with center on the y-axis passing through the origin, whereas if $c = 0$ we get $y = 0$, the x-axis. The v-curves are graphs of $v = \frac{x}{x^2+y^2} = d$. As above, if $d \neq 0$, we obtain a circle with center on the x-axis passing through the origin, and if $d = 0$ we obtain the y-axis. (Note, however, that the origin is not included on any u- or v-curve since the domain of T excludes (0,0).)

(b) First observe that $u^2+v^2 = \frac{y^2+x^2}{(x^2+y^2)^2} = \frac{1}{x^2+y^2}$. Thus $u = y \cdot \frac{1}{x^2+y^2} = y(u^2+v^2) \Rightarrow y = \frac{u}{u^2+v^2}$. $v = x \cdot \frac{1}{x^2+y^2} = x(u^2+v^2) \Rightarrow x = \frac{v}{u^2+v^2}$.

16. The u-surfaces are graphs of $u = x^2+y^2+z^2 = c \geq 0$, spheres of radius $\sqrt{c}$ with center at (0,0,0). The v-surfaces are graphs of $v = x^2+y^2-z^2 = d$, hyperboloids of one sheet if $d > 0$ or of two sheets if $d < 0$, or a cone with axis along the z-axis if $d = 0$. The w-surfaces are graphs of $w = x^2-y^2-z^2 = e$, hyperboloids of two sheets of $e > 0$ or of one sheet if $e < 0$, or a cone with axis along the x-axis if $e = 0$.

T is not one-to-one here. The points $(x,y,z) = (\pm 1, \pm 1, \pm 1)$ all have the same image $(u,v,w) = (3,1,-1)$.

EXERCISES 18.10, page 956

1. $\dfrac{\partial(x,y)}{\partial(u,v)} = \begin{vmatrix} \partial x/\partial u & \partial x/\partial v \\ \partial y/\partial u & \partial y/\partial v \end{vmatrix} = \dfrac{\partial x}{\partial u}\dfrac{\partial y}{\partial v} - \dfrac{\partial x}{\partial v}\dfrac{\partial y}{\partial u} = (2u)(2u)-(-2v)(2v) = 4(u^2+v^2).$

4. $\dfrac{\partial(x,y)}{\partial(u,v)} = \dfrac{v^2-u^2}{(u^2+v^2)^2} \cdot \dfrac{u^2-v^2}{(u^2+v^2)^2} - \left(\dfrac{-2uv}{(u^2+v^2)^2}\right)^2 = \dfrac{-u^4+2u^2v^2-v^4-4u^2v^2}{(u^2+v^2)^4} =$
$-\dfrac{(u^2+v^2)^2}{(u^2+v^2)^4} = -\dfrac{1}{(u^2+v^2)^2}.$

7. Each part of C, the boundary of R, gives rise to a part of K, the boundary of S, under the given transformation $x = u+v$, $y = 2v$. To determine K, and, therefore, S, we examine each part separately.

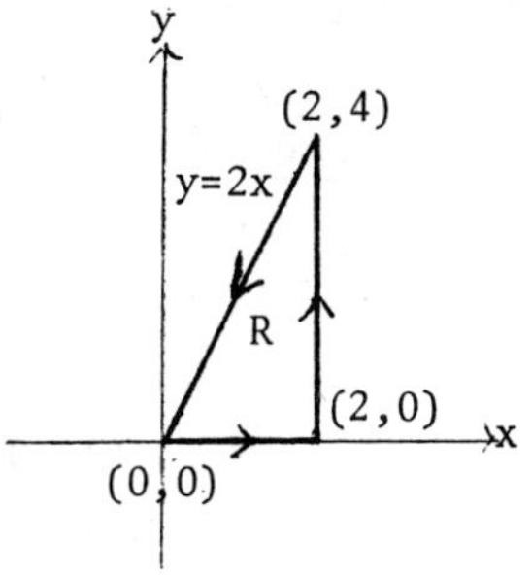

(i) The horizontal leg of C along the x-axis from (0,0) to (2,0) is given by $y = 0$, $0 \le x \le 2$.
$y = 0 \Longrightarrow v = 0$ and $u = x$, $0 \le x \le 2$. Thus, the corresponding portion of K is given by $v = 0$, $0 \le u \le 2$, i.e., it is on the u-axis between (0,0) and (2,0)

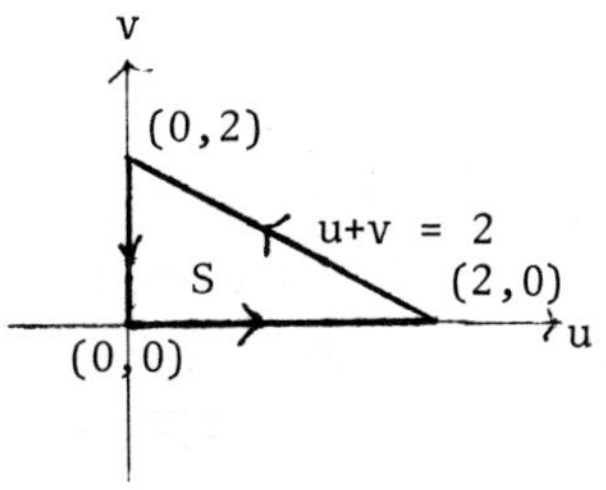

(ii) The vertical leg of C from (2,0) to (2,4) is given by $x = 2$, $0 \le y \le 4$,
$x = 2 \Longrightarrow 2 = u+v$. Since $v = y/2$, $0 \le y \le 4$ $\Longrightarrow 0 \le v \le 2$. Thus, this part of K is on the line $u+v = 2$, or $v = 2-u$. $v = 0 \Longrightarrow u = 2$ and $v = 2 \Longrightarrow u = 0$, i.e., the end-points of this segment are (0,2) and (2,0).

(iii) The 3rd part of C is given by $y = 2x$, $0 \le x \le 2$.
From the original transformation, $u = x-v$ and $v = y/2 \Longrightarrow u = x-y/2$. $y = 2x \Longrightarrow u = x-2x/2 = 0$ and $v = x$ satisfies $0 \le v \le 2$. Thus, this part of K is on the v-axis ($u = 0$) between (0,0) and (0,2).

These considerations produce S as the triangular region shown. Now, as (u,v) traverses K in the positive direction starting at the origin, the vertices (0,0), (2,0), (0,2), and (0,0) are encountered in that order (see arrows in sketch). The corresponding vertices on C are (0,0), (2,0), (2,4), and (0,0). Thus (x,y) traverses C once in the positive direction, and the "+" sign must be used in (18.47). Finally, we need $\dfrac{\partial(x,y)}{\partial(u,v)} = \dfrac{\partial x}{\partial u}\dfrac{\partial y}{\partial v} - \dfrac{\partial x}{\partial v}\dfrac{\partial y}{\partial u} = (1)(2)-(1)(0)=2.$

Thus $\displaystyle\iint_R (y-x)\,dx\,dy = \iint_S (2v-(u+v))\cdot 2\cdot dr\,du = \int_0^2\int_0^{2-u} (v-u)2\,dv\,du.$

10. We first note that the given transformation, $x = u^2-v^2$, $y = 2uv$, is not one-to-one. It will be so if we make the restrictions $u \geq 0$, $v \geq 0$. As before, we determine the boundary, K, of S and S itself.

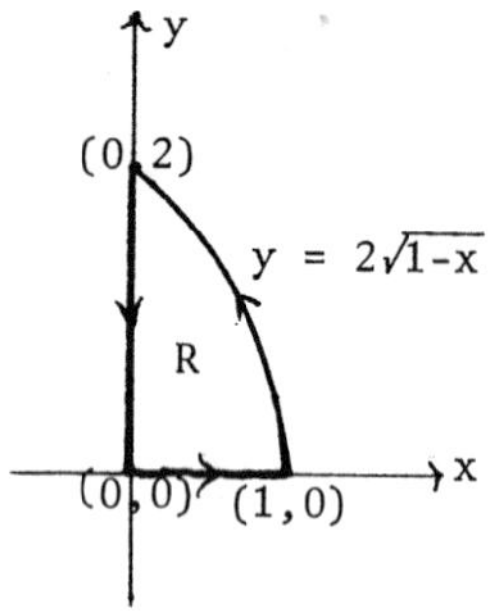

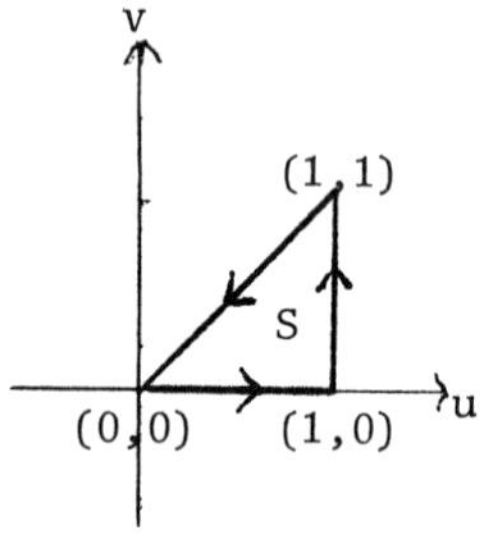

(i) The vertical leg of C is $x = 0$, $0 \leq y \leq 2$.

$x = 0 \Longrightarrow u^2 = v^2 \Longrightarrow u = \pm v$. $0 \leq y \leq 2 \Longrightarrow 0 \leq uv \leq 1 \Longrightarrow$ u and v have the same sign $\Longrightarrow u = v$ (i.e. we reject $u = -v$ from above) and in order to have $0 \leq uv \leq 1$, $u \geq 0$ and $v \geq 0$ it must be true that $0 \leq u \leq 1$ and $0 \leq v \leq 1$. Thus this part of K is on the line $u = v$ between (0,0) and (1,1).

(ii) The horizontal leg of C is $y = 0$, $0 \leq x \leq 1$.

$y = 0 \Longrightarrow 2uv = 0 \Longrightarrow u = 0$ or $v = 0$. $0 \leq x \leq 1 \Longrightarrow 0 \leq u^2-v^2 \leq 1$. If $u = 0$, then $0 \leq -v^2 \leq 1$ which is impossible on any v-interval. Thus $v = 0$ and $0 \leq u^2 \leq 1 \Longrightarrow 0 \leq u \leq 1$, and this part of K is on the u-axis ($v = 0$) between (0,0) and (1,0).

(iii) The curved part of C is $y = 2\sqrt{1-x}$, $0 \leq x \leq 1$.

Substituting from the original transformation, $y = 2\sqrt{1-x} \Longrightarrow 2uv = 2\sqrt{1-u^2+v^2} \Longrightarrow u^2v^2 = 1-u^2+v^2 \Longrightarrow u^2+u^2v^2-v^2-1 = 0 \Longrightarrow u^2(v^2+1)-(v^2+1) = 0 \Longrightarrow (u^2-1)(v^2+1) = 0 \Longrightarrow u^2=1 \Longrightarrow u = 1$. Next, $0 \leq x \leq 1 \Longrightarrow 0 \leq u^2-v^2 = 1-v^2 \leq 1 \Longrightarrow -1 \leq -v^2 \leq 0 \Longrightarrow 0 \leq v^2 \leq 1 \Longrightarrow 0 \leq v \leq 1$, and this part of K is on the line $u = 1$ between (1,0) and (1,1).

Now, as (u,v) traverses K in the positive direction through the vertices (0,0), (1,0), (1,1), and (0,0), in that order, the corresponding point (x,y) moves through the vertices (0,0), (1,0), (0,2), and (0,0) in that order. Thus (x,y) moves around C once in the positive direction so that the "+" sign in (18.47) must be used. Finally, the Jacobian, $\frac{\partial(x,y)}{\partial(u,v)} = 4(u^2+v^2)$ from #1 above.

Thus $\iint_R xy\,dxdy = \iint_S (u^2-v^2)(2uv)\cdot 4(u^2+v^2)\,dv\,du = \int_0^1 \int_0^u 8uv(u^4-v^4)\,dv\,du.$

13. $x = u/v$, $y = v \Longrightarrow u = xy$, $v = y$, and u and v are ≥ 0 since x and y are. Thus the parts of C given by $xy = 1$ and $xy = 2$ correspond to the vertical lines $u = 1$ and $u = 2$. Next, $y = x \Longrightarrow u = x^2$, $v = x \Longrightarrow v^2 = u \Longrightarrow v = \sqrt{u}$, $1 \leq u \leq 2$. Finally, $y = 2x \Longrightarrow u = 2x^2$, $v = 2x \Longrightarrow v = \sqrt{2u}$, $1 \leq u \leq 2$. As before, it is easy to verify that (x,y) moves around C in the positive direction as (u,v) does the same around K. Finally,

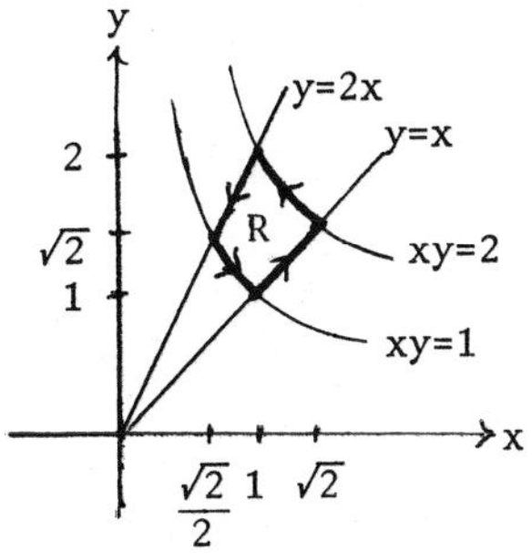

$\frac{\partial(x,y)}{\partial(u,v)} = \frac{\partial x}{\partial u}\frac{\partial y}{\partial v} - \frac{\partial x}{\partial v}\frac{\partial y}{\partial u} = \frac{1}{v}$. Thus $\iint_R (x^2+2y^2)\,dxdy$

$= \iint_S \left(\frac{u^2}{v^2} + 2v^2\right)\cdot\frac{1}{v}\,dv\,du = \int_1^2 \int_{\sqrt{u}}^{\sqrt{2u}} (u^2v^{-3}+2v)\,dv\,du$

$= \int_1^2 \left[-\frac{u^2v^{-2}}{2} + v^2\right]_{\sqrt{u}}^{\sqrt{2u}} du =$

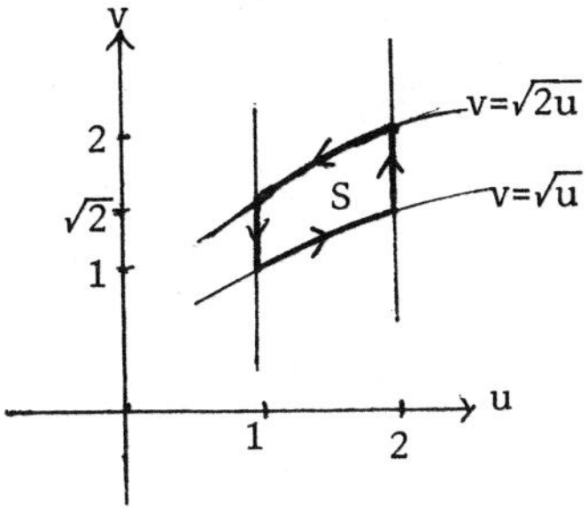

$\int_1^2 \left[\left(-\frac{u^2(2u)^{-1}}{2} + 2u\right) - \left(-\frac{u^2u^{-1}}{2} + u\right)\right] du = \int_1^2 \left(-\frac{u}{4} + 2u + \frac{u}{2} - u\right) du =$

$\frac{5}{4}\int_1^2 u\,du = \frac{15}{8}$.

16. $u = y/2$, $v = x-2y \Longrightarrow x = 4u+v$, $y = 2u$.

(i) The horizontal leg of C is $y = 0$, $0 \leq x \leq 4$.

$y = 0 \Longrightarrow u = 0$, $v = x$, $0 \leq x \leq 4 \Longrightarrow 0 \leq v \leq 4$, and this part of K is on the v-axis between (0,0) and (0,4).

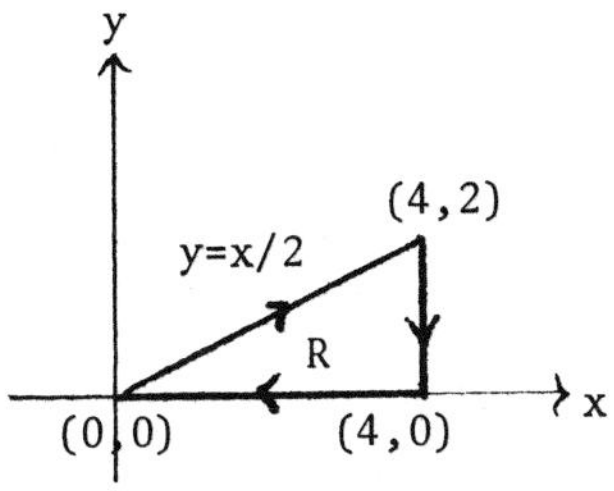

(ii) The vertical leg of C is $x = 4$, $0 \leq y \leq 2$.

$x = 4$, $y = 2u \Longrightarrow v = 4-2(2u) = 4-4u$.

$0 \leq y \leq 2 \Longrightarrow 0 \leq 2u \leq 2 \Longrightarrow 0 \leq u \leq 1$.

Thus, this part of K is on the line $v = 4-4u$ between (0,4) and (1,0).

(iii) Finally, the 3rd part of C is given by $y = x/2$, $0 \leq x \leq 4$.

$y = x/2 \Longrightarrow v = x-2(x/2) = 0$. $0 \leq x \leq 4 \Longrightarrow 0 \leq y \leq 2 \Longrightarrow 0 \leq u \leq 1$, and this part of K is on the u-axis between (0,0) and (1,0).

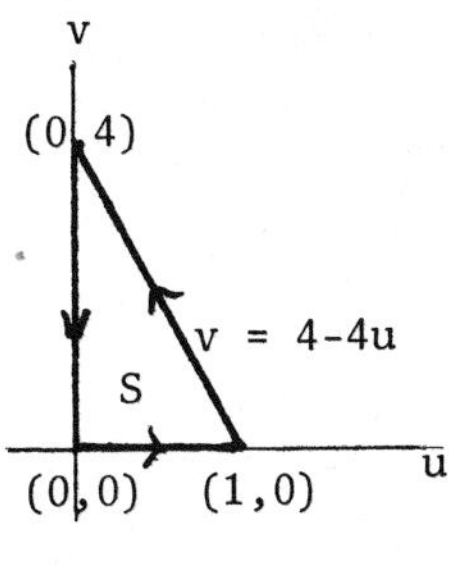

As (u,v) traverses K in the positive direction through the vertices (0,0), (1,0), (0,4), and (0,0) in that order, the corresponding point (x,y) moves through the vertices (0,0), (4,2), (4,0), and (0,0) in that order. Thus (x,y) traverses C once in the negative (clockwise) direction. Thus the "-"

sign is needed in (18.47). Finally, $\frac{\partial(x,y)}{\partial(u,v)} = (4)(0)-(1)(2) = -2$. Thus

$$\iint_S (\sqrt{x-2y} + \frac{y^2}{4})\,dx\,dy = -\iint_R (\sqrt{v}+u^2)(-2)\,dv\,du = 2\int_0^1 \int_0^{4-4u} (v^{1/2}+u^2)\,dv\,du =$$

$$2\int_0^1 [\frac{2}{3}v^{3/2} + u^2 v]_0^{4-4u} du = 2\int_0^1 [\frac{2}{3}(4-4u)^{3/2} + 4u^2 - 4u^3]\,du =$$

$$= 2[-\frac{1}{15}(4-4u)^{5/2} + \frac{4}{3}u^3 - u^4]_0^1 = 2(\frac{1}{3} + \frac{32}{15}) = \frac{74}{15}.$$

EXERCISES 18.11 (Review), page 958

4. If $r = (x^2+y^2+z^2)^{1/2}$, then $\underline{F} = \nabla(\frac{1}{r}) = -\frac{\underline{r}}{r^3}$ (See #19, Sec. 18.5). Thus the vector at a point (x,y,z) points toward the origin with magnitude $|\underline{F}| = \frac{r}{r^3} = \frac{1}{r^2}$ (e.g. it is like an inverse square force field).

7. $x = 1-t$, $y = t^2 \Longrightarrow dx = -dt$, $dy = 2t\,dt$, $0 \le t \le 2$ and $\int_C y^2dx + xy\,dy =$

$$\int_0^2 [t^4(-1) + (1-t)t^2(2t)]dt = \int_0^2 (2t^3-3t^4)dt = \frac{t^4}{2} - \frac{3}{5}t^5]_0^2 = 8 - \frac{96}{5} = -\frac{56}{5}.$$

10. $y = 2-2x^3 \Longrightarrow dy = -6x^2dx$ with x going from +1 to -1 to get the initial point (1,0) and terminal point (-1,4). Then $\int_C y^2dx + xy\,dy = \int_1^{-1} [(2-2x^3)^2$

$$+ x(2-2x^3)(-6x^2)]dx = \int_1^{-1} [4-20x^3+16x^6]dx = [4x-5x^4 + \frac{16}{7}x^7]_1^{-1} = -\frac{88}{7}.$$

13. The line segment from A(0,0,0) to B(2,4,8) has parametric equations $x = 2t$, $y = 4t$, $z = 8t$, $0 \le t \le 1$, and $\int_C x\,dx + (x+y)dy + (x+y+z)dz =$

$$\int_0^1 [(2t)2 + (6t)(4) + (14t)(8)]dt = \int_0^1 140\,t\,dt = 70.$$

16. $W = \oint_C \underline{F}\cdot d\underline{r} = \oint_C xy\ dx + y^2z\ dy + xz^2dz$ where $C = C_1 \cup C_2 \cup C_3 \cup C_4$, where $C_1: A_1 \to A_2$, $C_2: A_2 \to A_3$, etc.

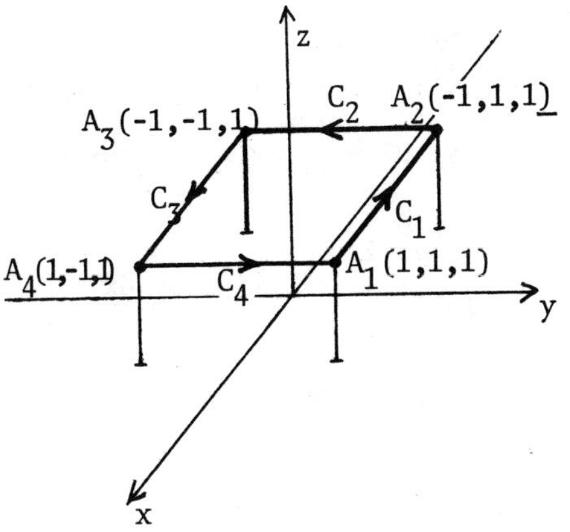

On C_1, $x = 1-t$, $y = 1$, $z = 1$, $0 \le t \le 2$, so $dy = dz = 0$ and

$$\int_{C_1} \underline{F}\cdot d\underline{r} = \int_0^2 (1-t)(-1)dt = \frac{(1-t)^2}{2}\Big]_0^2 = 0.$$

On C_2, $x = -1$, $y = 1-t$, $z = 1$, $0 \le t \le 2$, $dx = dz = 0$, and

$$\int_{C_2} = \int_0^2 (1-t)^2(1)(-1)dt = \frac{(1-t)^3}{3}\Big]_0^2 = -\frac{2}{3}.$$

On C_3, $x = t-1$, $y = -1$, $z = 1$, $0 \le t \le 2$, $dy = dz = 0$ and

$$\int_{C_3} = \int_0^2 (t-1)(-1)dt = -\frac{(t-1)^2}{2}\Big]_0^2 = 0.$$

On C_4, $x = 1$, $y = t-1$, $z = 1$, $0 \le t \le 2$, $dx = dz = 0$, and

$$\int_{C_4} = \int_0^2 (t-1)^2(1)dt = \frac{(t-1)^3}{3}\Big]_0^2 = \frac{2}{3}.$$

The sum of all 4 integrals is $W = 0$.

19. curl $\underline{F} = \underline{0}$ and by (18.41) we have independence of path. Now we want $f(x,y,z)$ such that $\nabla f = \underline{F}$, i.e. $f_x = 2xe^{2y}$, $f_y = 2x^2e^{2y} + 2y\cot z$, $f_z = -y^2\csc^2 z$. From the 1st equation, $f = x^2e^{2y} + g(y,z)$ and thus $f_y = 2x^2e^{2y} + g_y(y,z) = 2x^2e^{2y} + 2y\cot z$ (from the 2nd equation) $\Longrightarrow g_y(y,z) = 2y\cot z \Longrightarrow g(y,z) = y^2\cot z + h(z) \Longrightarrow f = x^2e^{2y} + y^2\cot z + h(z)$. This function does satisfy the 3rd equation if $h'(z) = 0$ or $h(z) = c$. Thus $f(x,y,z) = x^2e^{2y} + y^2\cot z + c$ is the potential function.

22. $M = xy$, $N = x^2+y^2 \Longrightarrow M_y = x$, $N_x = 2x$ and $\oint_C xy\ dx + (x^2+y^2)dy = \iint_R (2x-x)dA$

where R is the region $x^2+y^2 \le 1$. Using polar coordinates, we get

$$\int_0^{2\pi}\int_0^1 (r\cos\theta)r\ dr\ d\theta = \frac{1}{3}\int_0^{2\pi} \cos\theta\ d\theta = 0.$$

25. $z = x+y$ on $S \Rightarrow \sqrt{f_x^2 + f_y^2 + 1} = \sqrt{3}$. R is the triangular region shown, and is given by $\{(x,y): 0 \le x \le 1,\ 0 \le y \le 2-2x\}$

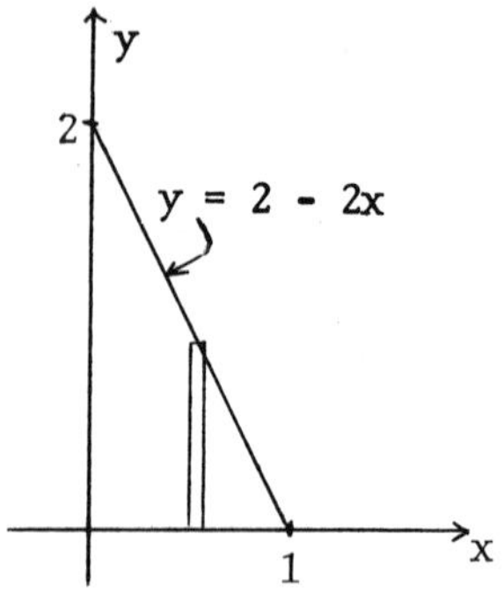

$$\iint_S xyz\ dS = \iint_R xy(x+y)\sqrt{3}\ dA =$$

$$\sqrt{3}\int_0^1\int_0^{2-2x}(x^2y+xy^2)\,dy\,dx =$$

$$\sqrt{3}\int_0^1\left[\frac{x^2(2-2x)^2}{2} + \frac{x(2-2x)^3}{3}\right]dx =$$

$$\sqrt{3}\int_0^1(2-2x)^2x\left(\frac{x}{2} + \frac{2-2x}{3}\right)dx =$$

$$\frac{\sqrt{3}}{6}\int_0^1(2-2x)^2x(4-x)\,dx = \frac{\sqrt{3}}{6}\int_0^1(-4x^4+24x^3-36x^2+16x)\,dx =$$

$$\frac{\sqrt{3}}{6}\left[-\frac{4}{5} + 6 - 12 + 8\right] = \frac{\sqrt{3}}{5}.$$

28. $\underline{F} = 2x\underline{i} + y\underline{j} - z\underline{k} \Rightarrow$ div $\underline{F} = 2+1-1 = 2$ and $\iiint_Q \text{div}\ \underline{F}\ dV = 2\iiint_Q dV =$ $2\cdot$volume of $Q = 2(2\cdot4\cdot6) = 96$ The box has 6 faces; we'll integrate over each one in turn and add. Our work is tabulated as follows: (the last column is the value of $\underline{F}\cdot\underline{n}$ on that particular face):

Face	Equation	n	$\underline{F}\cdot\underline{n}$	$\underline{F}\cdot\underline{n}\vert_{S_i}$
S_1	$z = -3$	$-\underline{k}$	z	-3
S_2	$z = 3$	$\underline{k}$	$-z$	-3
S_3	$x = -1$	$-\underline{i}$	$-2x$	2
S_4	$x = 1$	$\underline{i}$	$2x$	2
S_5	$y = -2$	$-\underline{j}$	$-y$	2
S_6	$y = 2$	$\underline{j}$	y	2

The region R_i corresponding to each face is a rectangle in a coordinate plane whose area is the same as that of the face. Now,

$$\iint_{S_1}\underline{F}\cdot\underline{n}\ dS = \iint_{S_2}\underline{F}\cdot\underline{n}\ dS = -3\cdot\text{Area of } S_1 = -3(2\cdot4) = -24.$$

$$\iint_{S_3}\underline{F}\cdot\underline{n}\ dS = \iint_{S_4}\underline{F}\cdot\underline{n}\ dS = 2\cdot\text{Area of } S_3 = 2(4\cdot6) = 48.$$

$$\iint_{S_5} \underline{F}\cdot\underline{n}\, dS = \iint_{S_6} \underline{F}\cdot\underline{n}\, dS = 2\cdot\text{Area of } S_5 = 2\cdot(2\cdot 6) = 24.$$

Summing, $\iint_S \underline{F}\cdot\underline{n}\, dS = -24-24+48+48+24+24 = 96.$

30. Both integrals work out to -3/2.

31. (c) $\left.\begin{array}{l} u = 2x + 5y \\ v = 3x - 4y \end{array}\right\} \Longrightarrow \left.\begin{array}{l} 4u = 8x+20y \\ 5v = 15x-20y \end{array}\right\}$ add to get $4u+5v = 23x$, or $x = (4u+5v)/23$.

Using this in either equation yields $y = (3u-2v)/23$.

(d) $\dfrac{\partial(x,y)}{\partial(u,v)} = \begin{vmatrix} \partial x/\partial u & \partial x/\partial v \\ \partial y/\partial u & \partial y/\partial v \end{vmatrix} = \begin{vmatrix} 4/23 & 5/23 \\ 3/23 & -2/23 \end{vmatrix} = -\dfrac{8}{23^2} - \dfrac{15}{23^2} = -\dfrac{23}{23^2} = -\dfrac{1}{23}$.

(e) $ax+by+c = 0 \Longrightarrow a\left(\dfrac{4u+5v}{23}\right) + b\left(\dfrac{3u-2v}{23}\right) + c = 0 \Longrightarrow (4a+3b)u + (5a-2b)v + 23c = 0.$

(f) $x^2+y^2 = a^2 \Longrightarrow \left(\dfrac{4u+5v}{23}\right)^2 + \left(\dfrac{3u-2v}{23}\right)^2 = a^2 \Longrightarrow (16u^2+40uv+25v^2) + (9u^2-12uv+4v^2) = (23a)^2 \Longrightarrow 25u^2 + 28uv + 29v^2 = (23a)^2.$

CHAPTER 19
DIFFERENTIAL EQUATIONS

EXERCISES 19.1, page 966

1. $y = C_1e^x + C_2e^{2x} \Longrightarrow y' = C_1e^x + 2C_2e^{2x} \Longrightarrow y'' = C_1e^x + 4C_2e^{2x}$. Substituting these into the left side of the equation: $y'' - 3y' + 2y = (C_1e^x + 4C_2e^{2x}) - (3C_1e^x + 6C_2e^{2x}) + (2C_1e^x + 2C_2e^{2x}) = C_1(1-3+2)e^x + C_2(4-6+2)e^{2x} = 0e^x + 0e^{2x}$ so that the end result is 0. Thus the given function satisfies the given equation.

4. $y = (C_1 + C_2x)e^{-2x} \Longrightarrow y' = (-2C_1 + C_2 - 2C_2x)e^{-2x} \Longrightarrow y'' = ((4C_1-4C_2) + 4C_2x)e^{-2x}$. Substituting: $y'' + 4y' + 4y = (4C_1 - 4C_2 + 4C_2x)e^{-2x} + (-8C_1 + 4C_2 - 8C_2x)e^{-2x} + (4C_1 + 4C_2x)e^{-2x} = [C_1(4-8+4) + C_2(-4+4) + C_2(4-8+4)x]e^{-2x}$ which again reduces to 0.

7. If $y^2-x^2-xy = C$, we differentiate implicitly to obtain: $2yy' - 2x - xy' - y = 0$ or $(2y-x)y'-(2x+y) = 0$ which yields the given equation when muliplied by -1.

10. $x^2+y^2 = c^2 \Longrightarrow 2x + 2yy' = 0 \Longrightarrow y' = -x/y$.

13. $y = C_1e^x + C_2xe^x \Longrightarrow y' = C_1e^x + C_2(x+1)e^x \Longrightarrow y'' = C_1e^x + C_2(x+2)e^x$. Noticing that $y'' + y = 2y'$, we obtain $y'' - 2y' + y = 0$.

16. Circles centered at (0,0) have equations $x^2+y^2 = c$. See #10 above.

19. $x\,dy - y\,dx = 0 \Longrightarrow \frac{dy}{y} - \frac{dx}{x} = 0 \Longrightarrow \ln|y| - \ln|x| = C \Longrightarrow \ln|\frac{y}{x}| = C \Longrightarrow |\frac{y}{x}| = e^C = C_1 \Longrightarrow |y| = C_1|x|$. Thus the solutions are $y = \pm C_1x$, or simply, $y = mx$ for any constant m.

22. $(xy-4x)dx + (x^2y+y)dy = 0 \Longrightarrow x(y-4)dx + y(x^2+1)dy = 0 \Longrightarrow \frac{x}{x^2+1}dx + \frac{y}{y-4}dy = 0 \Longrightarrow \frac{1}{2}\frac{2x}{x^2+1}dx + (1 + \frac{4}{y-4})dy = 0 \Longrightarrow \frac{1}{2}\ln(x^2+1) + y + 4\ln|y-4| = C \Longrightarrow \ln(x^2+1)^{1/2} + \ln(y-4)^4 + y = C$. Exponentiating both sides of the final equation yields an alternate form for the answer: $(x+1)^{1/2}(y-4)^4e^y = e^C = K$.

25. $e^{x+2y}dx - e^{2x-y}dy = 0 \Longrightarrow e^xe^{2y}dx - e^{2x}e^{-y}dy = 0 \Longrightarrow e^xe^{-2x}dx - e^{-2y}e^{-y}dy = 0$ (having multiplied by $e^{-2y}e^{-2x}$). This reduces to $e^{-x}dx - e^{-3y}dy = 0 \Longrightarrow -e^{-x} + \frac{1}{3}e^{-3y} = C \Longrightarrow e^{-3y} = 3C + 3e^{-x} \Longrightarrow -3y = \ln(3C + 3e^{-x}) \Longrightarrow y = -1/3\ln(3C + 3e^{-x})$.

28. $x^2y' - yx^2 = y \Rightarrow x^2dy - yx^2dx = ydx \Rightarrow x^2dy - y(x^2+1)dx = 0 \Rightarrow \frac{dy}{y} - \frac{x^2+1}{x^2}dx = 0 \Rightarrow \frac{dy}{y} - (1 + \frac{1}{x^2})dx = 0 \Rightarrow \ln|y| - (x - \frac{1}{x}) = C \Rightarrow \ln|y| = (x - \frac{1}{x}) + C \Rightarrow |y| = e^{x-1/x}e^C = C_1e^{x-1/x}$. Thus the solutions are $y = \pm C_1e^{(x-1/x)}$ or just $y = C_2e^{(x-1/x)}$.

31. $e^y \sin x\, dx - \cos^2 x\, dy = 0 \Rightarrow \frac{\sin x}{\cos^2 x}dx - e^{-y}dy = 0 \Rightarrow \sec x \tan x\, dx - e^{-y}dy = 0 \Rightarrow \sec x + e^{-y} = C$. Solving for y, $e^{-y} = C - \sec x \Rightarrow -y = \ln(C-\sec x) \Rightarrow y = -\ln(C-\sec x) = \ln(1/(C - \sec x)$.

34. $\sqrt{x}\, y' - \sqrt{y} = x\sqrt{y} \Rightarrow \sqrt{x}\, dy = \sqrt{y}(x+1)dx \Rightarrow y^{-1/2}dy = x^{-1/2}(x+1)dx = (x^{1/2} + x^{-1/2})dx \Rightarrow 2y^{1/2} = \frac{2}{3}x^{3/2} + 2x^{1/2} + C$. Using the condition that $y = 4$ when $x = 9$, we obtain the following equation for C: $2\cdot 4^{1/2} = \frac{2}{3}9^{3/2} + 2\cdot 9^{1/2} + C \Rightarrow 4 = \frac{2}{3}\cdot 27 + 2\cdot 3 + C \Rightarrow C = 4-18-6 = -20$. Thus $2y^{1/2} = \frac{2}{3}x^{3/2} + 2x^{1/2} - 20$ or $y = (\frac{1}{3}x^{3/2} + x^{1/2} - 10)^2$.

37. $(xy+x)dx + \sqrt{4+x^2}\, dy = 0 \Rightarrow x(y+1)dx + \sqrt{4+x^2}\, dy = 0 \Rightarrow \frac{x}{\sqrt{4+x^2}}dx + \frac{1}{y+1}dy = 0 \Rightarrow \sqrt{4+x^2} + \ln|y+1| = C$. Using the condition that $y = 1$ when $x = 0$, we obtain $\sqrt{4} + \ln|2| = C$, or $C = 2+\ln 2$. Thus, the solution can be written, for all $y \neq -1$, as $\sqrt{4+x^2} + \ln|y+1| = 2+\ln 2$. If $y > -1$, then $y+1 > 0$, $|y+1| = y+1$, and we can solve for y as follows: $\ln(y+1) = \ln 2+2 - \sqrt{4+x^2} \Rightarrow y+1 = e^{\ln 2+2-\sqrt{4+x^2}} \Rightarrow y = 2e^{2-\sqrt{4+x^2}} - 1$.

40. $\csc y\, dx - e^x dy = 0 \Rightarrow e^{-x}dx - \sin y\, dy = 0 \Rightarrow -e^{-x} + \cos y = C$. Setting $x = 0$, $y = 0$, we obtain $-e^0 + \cos 0 = C \Rightarrow -1+1 = C$ or $C = 0$. Thus the solution is $-e^{-x} + \cos y = 0$.

EXERCISES 19.2, page 970

NOTE: Each equation in this section is exact since $P_y = Q_x$ in each case. Thus every problem starts with the search for $F(x,y)$ such that $F_x = P$ and $F_y = Q$.

1. $F_x(x,y) = 2x+y$, $F_y(x,y) = 2y+x \Rightarrow F(x,y) = x^2 + xy + g(y)$ from the 1st equation. Then $F_y(x,y) = x + g'(y) = 2y + x \Rightarrow g'(y) = 2y \Rightarrow g(y) = y^2 + C_1$ and $F(x,y) = x^2+xy+y^2+C_1$. The solution is $x^2 + xy + y^2 = C$. (C_1 was absorbed into C.)

4. $F_x = 2x - 3x^2y$, $F_y = 1+2y-x^3$. From the 1st, $F = x^2 - x^3y + g(y) \Longrightarrow F_y = -x^3 + g'(y)$. Comparing with the 2nd, we get $g'(y) = 1+2y \Longrightarrow g(y) = y + y^2 + C_1$ and $F(x,y) = x^2 - x^3y + y + y^2 + C_1$. The solution is $x^2 - x^3y + y + y^2 = C$.

7. $F_x = e^{xy}(xy+1)$, $F_y = x^2e^{xy} + 2y$. From the 2nd, which is easier to integrate, $F = xe^{xy} + y^2 + g(x) \Longrightarrow F_x = x(ye^{xy}) + e^{xy} + g'(x)$ (Product rule in differentiating the 1st term). Comparing with the 1st equation we find that $g'(x) = 0$, so $g(x) = C_1$. Thus $F(x,y) = xe^{xy} + y^2 + C_1$ and the solution follows as given.

10. Rewriting the equation as $(y^2+2xy-x^2)dx + (x^2+2xy-y^2)dy = 0$, our conditions are $F_x = y^2 + 2xy - x^2$, $F_y = x^2 + 2xy - y^2$. From the 1st, $F = xy^2 - x^2y - \frac{x^3}{3} + g(y)$ and $F_y = 2xy + x^2 + g'(y)$. Comparing with the 2nd, $g'(y) = -y^2 \Longrightarrow g(y) = \frac{y^3}{3} + C_1$. Thus the solution is $xy^2 + x^2y - \frac{x^3}{3} - \frac{y^3}{3} = C$.

13. $F_x = x-yx^{-1} + \ln y$, $F_y = xy^{-1} - \ln x$. From the 1st, $F = \frac{x^2}{2} - y \ln x + x \ln y + g(y)$. (No $|x|$ needed in $\ln x$. From the $\ln x$ term in F_y, x must be positive.) Thus $F_y = -\ln x + \frac{x}{y} + g'(y)$ and $g'(y) = 0$, $g(y) = C_1$ and the solution is $\frac{x^2}{2} - y \ln x + x \ln y = C$.

16. After rewriting the equation in the $Pdx + Qdy = 0$ form, we get $F_x = P = 2xe^y + 3ye^x$, $F_y = Q = x^2e^y + 3e^x$, and $F = x^2e^y + 3ye^x + g(y) \Longrightarrow F_y = x^2e^y + 3e^x + g'(y)$ so that $g'(y) = 0$ and the solution is $x^2e^y + 3ye^x = C$. Using the condition that $y = 0$ when $x = 1$ we get the following equation for C: $1\cdot e^0 + 3\cdot 0\cdot e = C$ or $C = 1$. The particular solution desired is $x^2e^y+3ye^x = 1$.

EXERCISES 19.3, page 973

1. With the substitution $y = vx$, $dy = v\,dx + xdv$, the given equation $(x+3y)dx + x\,dy = 0$ becomes $(x+3vx)dx + x(v\,dx + x\,dv) = 0$. Multiplying it out and combining terms we get: $(x+4xv)dx + x^2dv = 0$ or $(1+4v)dx + x\,dv = 0$, where we divided through by x, assuming $x \neq 0$. This equation is separable and we solve it as in Section 1: $\frac{dx}{x} + \frac{dv}{1+4v} = 0 \Longrightarrow \ln|x| + \frac{1}{4}\ln|1+4v| = C \Longrightarrow$ $\ln|x|\left|1 + \frac{4y}{x}\right|^{1/4} = C \Longrightarrow |x|\left|1 + \frac{4y}{x}\right|^{1/4} = e^C = C_1 \Longrightarrow x(1 + \frac{4y}{x})^{1/4} = C_2$ or $x^4(1 + \frac{4y}{x}) = C_3$. (Note the many ways in which a solution can be expressed.)

4. With $y = vx$, $x^2dy + (y^2-xy)dx = 0$ becomes $x^2(v\,dx + x\,dv) + (v^2x^2 - x^2v)dx$ $= 0$ or $v^2x^2dx + x^3dv = 0 \Rightarrow \frac{dx}{x} + \frac{dv}{v^2} = 0 \Rightarrow \ln|x| - \frac{1}{v} = C \Rightarrow \ln|x| - \frac{x}{y} = C.$

7. $y' = \frac{x^3}{4x^3-3x^2y} = \frac{x}{4x-3y} \Rightarrow (4x-3y)dy - x\,dx = 0.$ With $y = vx$, we get $(4x-3vx)(v\,dx + x\,dv) - x\,dx = 0$ which reduces to $((4-3v)v - 1)dx + (4-3v)x\,dv$ $= 0$ or $\frac{dx}{x} - \frac{4-3v}{3v^2-4v+1}\,dv = 0.$ Since $3v^2-4v+1 = (3v-1)(v-1)$, we can express the dv coefficient in terms of partial fractions, $\frac{4-3v}{3v^2-4v+1} = \frac{A}{3v-1} + \frac{B}{v-1}$, and find (as in Sec. 10.4) $A = -9/2$, $B = 1/2$. Thus, integrating the last form of the differential equation above, we obtain:

$$\ln|x| + \frac{9}{2}\cdot\frac{1}{3}\ln|3v-1| - \frac{1}{2}\ln|v-1| = C$$

or

$$\ln\left|\frac{y}{x} - 1\right| - 3\ln\left|\frac{3y}{x} - 1\right| = 2\ln|x| + C_1$$

(which can be rewritten in many other forms, in particular, without absolute values if one assumes the appropriate quantities are positive).

10. As usual, $(x+y)dx = (x-y)dy$ becomes $(x+vx)dx = (x-vx)(v\,dx + x\,dv)$ or $[(1+v) - (1-v)v]dx = (1-v)x\,dv$ which reduces to $(1+v^2)dx = (1-v)x\,dv \Rightarrow$ $\frac{dx}{x} = \frac{1-v}{1+v^2}\,dv = (\frac{1}{1+v^2} - \frac{v}{1+v^2})dv.$ Thus if $x > 0$ we have $\ln x = (\tan^{-1}v - \frac{1}{2}\ln(1+v^2)) + C$ or $\ln x = \tan^{-1}(\frac{y}{x}) - \frac{1}{2}\ln(1 + \frac{y^2}{x^2}) + C.$

13. With $x = vy$, $dx = v\,dy + y\,dv$, the equation $3y\,dx + (x+2y)dy = 0$ becomes $3y(v\,dy + y\,dv) + (vy + 2y)dy = 0$ or $(3v+(v+2))dy + 3y\,dv = 0.$ Separating variables, we get $\frac{2}{y}\,dy + \frac{3}{2v+1}\,dv = 0$ which leads to $2\ln y + \frac{3}{2}\ln(2v+1) = C$ or $\ln y^2 + \frac{3}{2}\ln(\frac{2x}{y} + 1) = C.$

16. Both $P = x^2 + y^2$ and $Q = 2xy$ are homogeneous of degree 2. Also, $P_y = Q_x = 2y$. Thus the equation is both homogeneous and exact. It is usually easier to solve it as an exact equation. Thus we seek $F(x,y)$ such that $F_x = x^2+y^2$, $F_y = 2xy$. From the 1st, $F = \frac{x^3}{3} + xy^2 + g(y) \Rightarrow F_y = 2xy + g'(y)$, and so $g'(y) = 0$, $g(y) = C_1$, and the solution is $\frac{x^3}{3} + xy^2 = C.$

EXERCISES 19.4, page 976

1. $y' + 2y = e^{2x}$ is in the required form (i.e. coefficient of y' is 1) with $P(x) = 2$. Thus $\int P = 2x$ and the integrating factor is $e^{\int P} = e^{2x}$. Multiplying the equation by this factor we obtain: $e^{2x}y' + 2e^{2x}y = e^{4x} \Rightarrow \frac{d}{dx}(e^{2x}y) =$

$e^{4x} \Rightarrow e^{2x}y = \frac{1}{4}e^{4x} + C \Rightarrow y = \frac{1}{4}e^{2x} + Ce^{-2x}$.

4. Here $P(x) = \cot x = \cos x/\sin x$, and $\int P = \ln|\sin x|$ and $e^{\int P} = |\sin x| = \pm \sin x$. No matter which sign is valid, when we multiply the original equation, $y' + (\cot x)y = \csc x$, by the integrating factor, we obtain $(\sin x)y' + (\cos x)y = 1 \Rightarrow \frac{d}{dx}((\sin x)y) = 1 \Rightarrow (\sin x)y = x+C \Rightarrow y = (x+C)/\sin x = (x+C)\csc x$.

NOTE: Because of the lack of significance of the + or - sign when multiplying an equation by the integrating factor (#4 above, and Example 1), we will ignore the absolute value symbols when $\int P$ is a natural logarithm.

7. First divide by dx to obtain $x^2y' + 2xy = e^x$, and then rewrite in the form $y' + \frac{2}{x}y = \frac{e^x}{x^2}$. Thus $P(x) = \frac{2}{x}$, and the integrating factor is $e^{\int P} = e^{2 \ln x} = e^{\ln x^2} = x^2$. Multiplying by the integrating factor yields $x^2y' + 2xy = e^x \Rightarrow \frac{d}{dx}(x^2y) = e^x \Rightarrow x^2y = e^x+C \Rightarrow y = (e^x+C)/x^2$.

10. Here, $P(x) = \tan x$, and the integrating factor is $e^{\int P} = e^{\int \tan x} = e^{\int \sin x/\cos x} = e^{-\ln(\cos x)} = e^{\ln(\sec x)} = \sec x$. Multiplying by $\sec x$, we get: $(\sec x)y' + (\sec x \tan x)y = \sec x \sin x = \frac{\sin x}{\cos x} \Rightarrow \frac{d}{dx}(\sec x)y = \frac{\sin x}{\cos x} \Rightarrow (\sec x)y = \ln|\sec x| + C \Rightarrow y = (\ln|\sec x| + C)/\sec x = (\ln|\sec x| + C)\cos x$.

13. Rewriting the equation as $x^2\cos x + y - xy' = 0$ or $xy' - y = x^2\cos x$, we convert it into the required form (coefficient of y' to be 1) to obtain $y' - \frac{1}{x}y = x \cos x$. Thus $P(x) = -1/x$, $\int P = -\ln x = \ln(1/x)$, and $e^{\int P} = 1/x$. Multiplying by this we have $\frac{1}{x}y' - \frac{1}{x^2}y = \cos x \Rightarrow \frac{d}{dx}(\frac{1}{x}y) = \cos x \Rightarrow \frac{1}{x}y = \sin x + C \Rightarrow y = x \sin x + Cx$.

16. Rewriting the equation as $y' + \frac{5}{x+4} = \frac{x^2+8x+16}{x+4} = \frac{(x+4)^2}{x+4} = x+4$, we have $P(x) = \frac{5}{x+4}$, $\int P = 5 \ln(x+4) = \ln(x+4)^5$ and $e^{\int P} = (x+4)^5$. Multiplying, we get $(x+4)^5y' + 5(x+4)^4y = (x+4)^6 \Rightarrow \frac{d}{dx}((x+4)^5y) = (x+4)^6 \Rightarrow (x+4)^5y = \frac{(x+4)^7}{7} + C \Rightarrow y = \frac{(x+4)^2}{7} + \frac{C}{(x+4)^5}$.

19. The given equation can be rewritten as $(2xy - x^2y^2) + (1+x^2)y' = 0$ and then $y' + \frac{2x}{1+x^2}y = \frac{x^2}{1+x^2}y^2$. This is the Bernoulli equation with $n = 2$. So, we let

$w = y^{1-n} = y^{-1}$ or $y = w^{-1}$. Then $y' = -w^{-2}w'$, $y^2 = w^{-2}$. Making these substitutions yields $-w^{-2}w' + \frac{2x}{1+x^2}w^{-1} = \frac{x^2}{1+x^2}w^{-2}$ which, when multiplied by $-w^{-2}$ becomes the linear equation $w' - \frac{2x}{1+x^2}w = -\frac{x^2}{1+x^2}$. Here $P(x) = -\frac{2x}{1+x^2}$, $\int P = -\ln(1+x^2) = \ln(\frac{1}{1+x^2})$, $e^{\int P} = \frac{1}{1+x^2}$. As usual, we get $\frac{1}{1+x^2}w' = \frac{2x}{(1+x^2)^2}w$ $= -\frac{x^2}{(1+x^2)^2} \Longrightarrow \frac{d}{dx}(\frac{1}{1+x^2}w) = \frac{-x^2}{(1+x^2)^2} = -\frac{1}{1+x^2} + \frac{1}{(1+x^2)^2}$ (The partial fractions as in Sec. 10.4). Thus $\frac{1}{1+x^2}w = -\tan^{-1}x + \int\frac{1}{(1+x^2)^2}dx + C$. For this integral, let $x = \tan\theta$ to obtain $\int\frac{\sec^2\theta}{\sec^4\theta}d\theta = \int\cos^2\theta\, d\theta = \frac{1}{2}(\theta + \frac{1}{2}\sin 2\theta)$ $= \frac{1}{2}(\theta + \sin\theta\cos\theta) = \frac{1}{2}(\tan^{-1}x + \frac{x}{\sqrt{x^2+1}}\cdot\frac{1}{\sqrt{x^2+1}}) = \frac{1}{2}\tan^{-1}x + \frac{x}{x^2+1}$.

Using this in the w equation and solving for $w = \frac{1}{y}$, we finally get

$$\frac{1}{y} = -\frac{1}{2}(1+x^2)\tan^{-1}x + \frac{x}{2} + C(1+x^2).$$

22. $P(x) = -5$, $\int P = -5x$, $e^{\int P} = e^{-5x}$. Multiplying by this factor we get $e^{-5x}y'$ $-e^{-5x}\,5y = \frac{d}{dx}(e^{-5x}y) = 1$. Thus $e^{-5x}y = x + C \Longrightarrow y = (x+C)e^{5x}$.

25. $P(x) = 3x^2 \Longrightarrow e^{\int P} = e^{x^3}$ and using this factor, the equation becomes $\frac{d}{dx}(e^{x^3}y) = x^2e^{x^3} + 1 \Longrightarrow e^{x^3}y = \frac{1}{3}e^{x^3} + x + C \Longrightarrow y = \frac{1}{3} + (x+C)e^{-x^3}$.

28. $P(x) = 2 \Longrightarrow e^{\int P} = e^{2x}$ and the equation becomes $\frac{d}{dx}(e^{2x}y) = e^{-x} \Longrightarrow e^{2x}y =$ $-e^{-x} + C \Longrightarrow y = -e^{-3x} + Ce^{-2x}$. Setting $x = 0$, $y = 2$ we obtain $2 = -1 + C$ or $C = 3$. Thus the desired solution is $y = -e^{-3x} + 3e^{-2x}$.

EXERCISES 19.5, page 982

1. $x^2-y^2 = c \Longrightarrow 2x - 2yy' = 0 \Longrightarrow y' = x/y$. Thus, on the orthogonal trajectories $y' = -y/x$ or $y' + \frac{1}{x}y = 0$. This is linear with integrating factor $e^{\int\frac{1}{x}} = e^{\ln x} = x$. As in Sec. 4, we obtain by multiplying: $xy' + y = \frac{d}{dx}(xy)$ $= 0 \Longrightarrow xy = k \Longrightarrow y = k/x$, the equations of the orthogonal trajectories.

4. $y = cx^2 \Longrightarrow c = y/x^2$ and differentiation yields $y' = 2cx = 2(y/x^2)x = \frac{2y}{x}$. So, on the orthogonal trajectories $y' = -\frac{x}{2y} \Longrightarrow 2y\,dy + x\,dx = 0$ which is exact ($P = x$, $Q = 2y$ and $P_y = Q_x = 0$). As in Sec. 2 we seek $F(x,y)$ such that $F_x = x$, $F_y = 2y$ and easily find $F(x,y) = \frac{x^2}{2} + y^2 + C$, so that the

solution is $\frac{x^2}{2} + y^2 = k$. The orthogonal trajectories are ellipses.

7. $x^2 + y^2 = 2cx \Longrightarrow 2x + 2yy' = 2c$. Solving the original equation for $2c = \frac{x^2+y^2}{x}$ we obtain $2x + 2yy' = \frac{x^2+y^2}{x}$. Solving for this y': $2yy' = x + \frac{y^2}{x} - 2x = \frac{y^2-x^2}{x} \Longrightarrow y' = \frac{y^2-x^2}{2xy}$ so that on the orthgonal trajectories $y' = \frac{2xy}{x^2-y^2}$ or $2xy\,dx - (x^2-y^2)dy = 0$, which is homogeneous. The substitution $x = vy$, $dx = v\,dy + y\,dv$ produces $2vy^2(v\,dy + y\,dv) + (y^2 - v^2y^2)\,dy = 0 \Longrightarrow 2vy^3dv + y^2(1+v^2)dy = 0 \Longrightarrow \frac{2v}{1+v^2}dv + \frac{1}{y}dy = 0 \Longrightarrow \ln(1+v^2) + \ln|y| = C_1 \Longrightarrow y(1+v^2) = \pm e^{C_1} \Longrightarrow y(1 + \frac{x^2}{y^2}) = k$ or $x^2 + y^2 = ky$. Thus the original family consisted of circles centered on the x-axis. The orthogonal trajectories are also circles, but centered on the y-axis.

10. $RQ' + \frac{1}{C}Q = E \Longrightarrow Q' + \frac{1}{RC}Q = \frac{E}{R}$, linear with integrating factor $e^{\int P} = e^{\int\frac{1}{RC}dt} = e^{\frac{1}{RC}dt}$. Multiplying as usual: $e^{\frac{1}{RC}t}Q' + e^{\frac{1}{RC}t}\frac{1}{RC}Q = \frac{d}{dt}(e^{\frac{1}{RC}t}Q) = \frac{E}{R}e^{\frac{1}{RC}t} \Longrightarrow e^{\frac{1}{RC}t}Q = CE\,e^{\frac{t}{RC}} + K \Longrightarrow Q(t) = CE + Ke^{-t/RC}$. Since $Q(0) = 0$ we obtain $0 = CE + K$ or $K = -CE$. Thus $Q(t) = CE(1 - e^{-t/RC})$.

13. Using the notation of Example 1, the downward force is now $mg - kv^2$ and, with $c = k/m$, the differential equation is $\frac{dv}{g-cv^2} = dt$ or $\frac{dv}{a^2-v^2} = c\,dt$ where $a^2 = g/c$. Using partial fractions, $\frac{1}{a^2-v^2} = \frac{1}{2a}(\frac{1}{a+v} + \frac{1}{a-v})$, and the equation can be integrated. We obtain $\frac{1}{2a}[\ln(a+v) - \ln(a-v)] = ct + K$ or $\frac{1}{2a}\ln\frac{a+v}{a-v} = ct + K$. Using the initial condition that $v = 0$ when $t = 0$, $K = 0$ results. Thus $\ln\frac{a+v}{a-v} = 2act \Longrightarrow \frac{a+v}{a-v} = e^{2act}$ For the moment, let $b = 2ac$ $(= 2\sqrt{gc})$.

Solving for this last equation for v we obtain:

$$\frac{ds}{dt} = v = a\frac{e^{bt}-1}{e^{bt}+1} = a(\frac{e^{bt}}{e^{bt}+a} - \frac{1}{e^{bt}+1}) = a(\frac{e^{bt}}{e^{bt}+1} - \frac{e^{-bt}}{1+e^{-bt}})$$

and, integrating this:

$$s = \frac{a}{b}[\ln(e^{bt}+1) + \ln(1+e^{-bt})] + C$$

$$= \frac{1}{2c}[\ln(e^{bt}+1) + \ln(\frac{e^{bt}+1}{e^{bt}})] + C$$

$$= \frac{1}{2c}[2\ln(e^{bt}+1) - bt] + C$$

$$= \frac{1}{c}[\ln(e^{2\sqrt{gc}\,t}+1) - \sqrt{gc}\,t] + C$$

Since s = 0 when t = 0, it follows that $C = \frac{-\ln 2}{c}$.

EXERCISES 19.6, page 989

1. The auxiliary equation is $m^2-5m+6 = 0$ or $(m-2)(m-3) = 0$ with solutions m = 2 and 3. Thus the general solution is $y = C_1e^{2x} + C_2e^{3x}$.

4. The auxiliary equation is $m^2+6m+8 = 0$ or $(m+2)(m+4) = 0$ with solutions m = -2 and -4. Thus the general solution is $y = C_1e^{-2x} + C_2e^{-4x}$.

7. $m^2-4m+1 = 0 \Longrightarrow m = \frac{4 \pm \sqrt{16-4}}{2} = \frac{4 \pm 2\sqrt{3}}{2} = 2 \pm \sqrt{3}$. Thus $y = C_1e^{(2+\sqrt{3})x} + C_2e^{(2-\sqrt{3})x}$.

10. $4m^2 + 20m + 25 = (2m+5)^2 = 0 \Longrightarrow$ there is a double root, m = -5/2. Thus $y = (C_1 + C_2x)e^{-\frac{5}{2}x}$.

13. Divide by 9 to get $y'' - \frac{24}{9}y' + \frac{16}{9}y = 0$. The auxiliary equation is $m^2 - \frac{8}{3}m + \frac{16}{9} = (m - \frac{4}{3})^2 = 0$ with a double root, $m = \frac{4}{3}$. Thus $y = (C_1+C_2x)e^{(4/3)x}$.

16. Divide by 2 to get $y'' + \frac{7}{2}y' = 0$. The auxiliary equation is $m^2 + \frac{7}{2}m = m(m + \frac{7}{2}) = 0$ with solutions m = 0 and $m = -\frac{7}{2}$. Thus $y = C_1e^{0x} + C_2e^{-(7/2)x} = C_1 + C_2e^{-(7/2)x}$.

19. $m^2-4m+13 = 0 \Longrightarrow m = \frac{4 \pm \sqrt{16-52}}{2} = \frac{4 \pm \sqrt{-36}}{2} = \frac{4 \pm 6i}{2} = 2 \pm 3i$, complex roots. Thus $y = e^{2x}(C_1\cos 3x + C_2 \sin 3x)$.

22. $m^2+2m+6 = 0 \Longrightarrow m = \frac{-2 \pm \sqrt{4-24}}{2} = -1 \pm \sqrt{5}i$ and $y = e^{-x}(C_1\cos \sqrt{5}x + C_2\sin \sqrt{5}x)$.

25. First, we find the general solution. $m^2+1 = 0 \Longrightarrow m = \pm i \Longrightarrow y = C_1\cos x + C_2\sin x$. Next, we use the given initial conditions to determine C_1 and C_2. From y we obtain $y' = -C_1\sin x + C_2\cos x$. Setting x = 0, we see that $y(0) = C_1$, $y'(0) = C_2$. Thus, in order to have y(0) = 1, y'(0) = 2, we must take $C_1 = 1$, $C_2 = 2$, and the desired solution is y = cos x + 2 sin x.

28. $m^2+5 = 0 \Longrightarrow m = \pm\sqrt{5}\,i$ and the general solution is $y = C_1\cos \sqrt{5}x + C_2\sin \sqrt{5}x$. From this, $y' = -\sqrt{5}C_1\sin \sqrt{5}x + \sqrt{5}C_2\cos \sqrt{5}x$. Setting x = 0 we find $y(0) = C_1$,

$y'(0) = \sqrt{5}C_2$. The initial conditions $y(0) = 4$, $y'(0) = 2$ will then be satisfied by choosing $C_1 = 4$, $C_2 = 2/\sqrt{5}$, and the desired solution is $y = 4\cos\sqrt{5}x + (2/\sqrt{5})\sin\sqrt{5}x$.

EXERCISES 19.7, page 995

1. Let $y_1 = \cos x$, $y_2 = \sin x$--the basic solutions of the complementary equation $y'' + y = 0$. Then we look for a particular solution of the given equation, $y'' + y = \tan x$, of the form $y_p = uy_1 + vy_2$. The system of equations for u' and v' is:

$$\left.\begin{aligned} u'\cos x + v'\sin x &= 0 \\ -u'\sin x + v'\cos x &= \tan x \end{aligned}\right\} \Longrightarrow \left\{\begin{aligned} u'\cos x\sin x + v'\sin^2 x &= 0 \\ -u'\cos x\sin x + v'\cos^2 x &= \sin x \end{aligned}\right.$$

Adding yields $v' = \sin x$. Substituting this into the 1st equation produces $u'\cos x + \sin^2 x = 0$ or $u' = -\dfrac{\sin^2 x}{\cos x} = \dfrac{-1+\cos^2 x}{\cos x} = -\sec x + \cos x$. Integrating these v' and u' results we get $u = -\ln|\sec x + \tan x| + \sin x$, $v = -\cos x$. Thus $y_p = uy_1 + vy_2 = -(\cos x)\ln|\sec x + \tan x|$ + ~~sin x cos x~~ - ~~cos x sin x~~.
The general solution is $y = C_1\cos x + C_2\sin x - (\cos x)\ln|\sec x + \tan x|$.

4. We first solve the complementary equation, $y'' + 3y' = 0$. The auxiliary equation is $m^2+3m = m(m+3) = 0 \Longrightarrow m = 0$ and -3. Thus the homogeneous equation has basic solutions $y_1 = e^{0x} = 1$, $y_2 = e^{-3x}$, and the system of equations for u' and v', so that $y_p = uy_1 + vy_2$ will be a solution of $y'' + 3y' = e^{-3x}$, is

$$\left.\begin{aligned} u'\cdot 1 + v'e^{-3x} &= 0 \\ -3v'e^{-3x} &= e^{-3x} \end{aligned}\right\} \Longrightarrow -3v' = 1 \Longrightarrow v' = -1/3.$$

Using this in the 1st equation, we get $u' = \frac{1}{3}e^{-3x}$. Integrating u' and v' we get $u = -\frac{1}{9}e^{-3x}$, $v = -\frac{x}{3}$ so that $y_p = uy_1 + vy_2 = -\frac{1}{9}e^{-3x} - \frac{x}{3}e^{-3x}$ and the general solution is $y = C_1 + C_2e^{-3x} + y_p = C_1 + C_2'e^{-3x} - \frac{x}{3}e^{-3x}$ where $C_2' = C_2 - \frac{1}{9}$.

7. First solving the complementary equation $y''-9y = 0$, we obtain $y_1 = e^{3x}$, $y_2 = e^{-3x}$ as basic solutions ($m^2-9 = 0 \Longrightarrow m = \pm 3$). The system for u' and v' is

$$\left.\begin{array}{l} u'\,e^{3x} + v'\,e^{-3x} = 0 \\ 3u'\,e^{3x} - 3v'\,e^{-3x} = e^{3x} \end{array}\right\} \Longrightarrow 6u'e^{3x} = e^{3x} \Longrightarrow u' = 1/6$$

(We added 3 times the 1st to the 2nd.) Using this in the 1st equation, $v'e^{-3x} = -\frac{1}{6}e^{3x} \Longrightarrow v' = -\frac{1}{6}e^{6x}$. Integrating, $u = \frac{x}{6}$, $v = -\frac{1}{36}e^{6x}$ and $y_p = uy_1 + vy_2 = \frac{x}{6}e^{3x} - \frac{1}{36}e^{3x}$, and the general solution is $y = C_1e^{3x} + \frac{x}{6}e^{3x} + C_2e^{-3x}$. (The constant $-\frac{1}{36}$ was absorbed into the arbitrary constant C_1.)

10. The complementary equation $\frac{d^2y}{dx^2} - \frac{dy}{dx} = 0$ has basic solutions $y_1 = 1$, $y_2 = e^x$ ($m^2-m = 0 \Longrightarrow m = 0,1$). The system for u' and v' is

$$\left.\begin{array}{l} u' + v'\,e^{x} = 0 \\ v'\,e^{x} = x+1 \end{array}\right\} \Longrightarrow u' = -x-1 \Longrightarrow v' = (x+1)e^{-x}.$$

Integrating, $u = -\frac{x^2}{2} - x$, $v = -xe^{-x} - 2e^{-x}$ (using $\int xe^{-x}\,dx = -(x+1)e^{-x} + c$), and $y_p = uy_1 + vy_2 = -\frac{x^2}{2} - x - (xe^{-x} + 2e^{-x})e^x = -\frac{x^2}{2} - 2x - 2$, and the general solution is $y = C_1 + C_2e^x - \frac{x^2}{2} - x$. (The constant -2 is actually a solution of the homogeneous equation, and it was absorbed into C_1.)

13. The complementary equation, $y''+2y' = 0$, has auxiliary equation $m^2+2m = 0$ with solutions $m = 0,-2$. The right side of the given equation is cos 2x, and since $0+2i = 2i$ is not a solution of the auxiliary equation, by rule (3) there is a particular solution of the form $y_p = A\cos 2x + B\sin 2x$. We now substitute this into the left side of the given equation, set the result equal to the given right side, and solve for A and B. Differentiating y_p, we obtain $y_p' = -2A\sin 2x + 2B\cos 2x$, $y_p'' = -4A\cos 2x - 4B\sin 2x$. Thus $y_p'' + 2y_p' = (-4A + 4B)\cos 2x + (-4A-4B)\sin 2x$. This equals cos 2x if:

$$\left.\begin{array}{l} -4A + 4B = 1 \\ -4A - 4B = 0 \end{array}\right\} \Longrightarrow A = -1/8,\ B = 1/8.$$

Thus $y_p = (1/8)(\sin 2x - \cos 2x)$ and the general solution is $y = C_1 + C_2e^{-2x} + (1/8)(\sin 2x - \cos 2x)$.

16. The associated homogeneous equation, $y'' + 3y' - 4y = 0$, has auxiliary equation $m^2+3m-4 = (m+4)(m-1) = 0$ with solutions $m = 1,-4$. The right side of the given equation is $x\,e^{-x}$ for which $n = -1$ is not a solution of the auxiliary equation. Thus, by rule (2), there is a particular solution of the

form $y_p = (A+Bx)e^{-x}$ so that $y_p' = ((-A+B)-Bx)e^{-x}$, $y_p'' = ((A-2B)+Bx)e^{-x}$. Now we substitute these into the differential equation to determine A and B. $y_p'' + 3y_p' - 4y_p = \{[(A-2B)-3A+3B-4A] + [B-3B-4B]\}e^{-x} = [(-6A+B)-6Bx]e^{-x}$. This will equal the right side, xe^{-x}, if $-6A+B = 0$, $-6B = 1$. Thus $B = -1/6$, and $A = \frac{1}{6}B = -1/36$. Thus $y_p = (-\frac{1}{36} - \frac{1}{6}x)e^{-x}$ and the general solution is $y = C_1e^x + C_2e^{-4x} - (\frac{1}{36} + \frac{1}{6}x)e^{-x}$.

EXERCISES 19.8, page 997

1. To solve $y'' + y = 0$ in terms of power series, we assume $y = \sum_{n=0}^{\infty} a_n x^n$ so that $y' = \sum_{n=1}^{\infty} na_n x^{n-1}$, $y'' = \sum_{n=2}^{\infty} n(n-1)a_n x^{n-2}$, and, substituting these into the differential equation, we obtain: $y'' + y = \sum_{n=2}^{\infty} n(n-1)a_n x^{n-2} + \sum_{n=0}^{\infty} a_n x^n = 0$.

In the 1st series, we replace n-2 by n to have the same power of x in each series. To do this, first let $k = n-2$, so that $n = k+2$, $n-1 = k+1$, and $n = 2$ corresponds to $k = 0$. Then replace the k by n. We obtain

$$\sum_{n=0}^{\infty} (n+2)(n+1)a_{n+2}x^n + \sum_{n=0}^{\infty} a_n x^n = 0$$

$$\sum_{n=0}^{\infty} [(n+2)(n+1)a_{n+2} + a_n]x^n = 0.$$

We now equate all coefficients to 0. (You can think of the 0 on the right side as a power series all of whose coefficients are 0.) The resulting formula is $(n+2)(n+1)a_{n+2} + a_n = 0$, or $a_{n+2} = -\frac{a_n}{(n+2)(n+1)}$. This leads to

$a_2 = -\frac{a_o}{2}$	$a_3 = -\frac{a_1}{3\cdot 2} = -\frac{a_1}{3!}$
$a_4 = -\frac{a_2}{4\cdot 3} = \frac{a_o}{4!}$	$a_5 = -\frac{a_3}{5\cdot 4} = \frac{a_1}{5!}$
$a_6 = -\frac{a_4}{6\cdot 5} = -\frac{a_o}{6!}$	$a_7 = -\frac{a_5}{7\cdot 6} = -\frac{a_1}{7!}$
.	.
.	.
.	.
$a_{2k} = \frac{(-1)^k a_o}{(2k)!}$	$a_{2k+1} = \frac{(-1)^k a_1}{(2k+1)!}$

Thus the series for the solution, y, breaks into 2 parts: one consisting of the even powers of x with coefficients a_{2k}, the other consisting of odd powers of x with coefficients a_{2k+1}. The 2 coefficients a_o and a_1 are arbitrary. (In an initial value problem a_o would be y(0), and a_1 is y'(0).) Thus

$$y = a_o \sum_{k=0}^{\infty} \frac{(-1)^k}{(2k)!} x^{2k} + a_1 \sum_{k=0}^{\infty} \frac{(-1)^k}{(2k+1)!} x^{2k+1}$$

(You should recognize these series as the Maclaurin series for cos x and sin x, respectively, so that we have the same general solution of y'' + y = 0 that would be obtained by the method of Sec. 19.6.)

4. Again let $y = \sum_{n=0}^{\infty} a_n x^n$, $y' = \sum_{n=1}^{\infty} na_n x^{n-1}$, $y'' = \sum_{n=2}^{\infty} n(n-1)a_n x^{n-2}$ and substitute.

$$y'' + 2xy' + y = \sum_{n=2}^{\infty} n(n-1)a_n x^{n-2} + 2x \sum_{n=1}^{\infty} na_n x^{n-1} + \sum_{n=0}^{\infty} a_n x^n = 0.$$

In the 1st series replace n-2 by n (See #1 above) and take the x inside the 2nd series, changing the exponent of x to n. This yields

$$\sum_{n=0}^{\infty} (n+2)(n+1)a_{n+2}x^n + 2 \sum_{n=1}^{\infty} na_n x^n + \sum_{n=0}^{\infty} a_n x^n = 0$$

Now, we'll take the n = 0 term out of the 1st and 3rd series so that all remaining series begin the summation at n = 1. Combining these we get:

$$(2\cdot 1a_2 + a_o) + \sum_{n=1}^{\infty} [(n+2)(n+1)a_{n+2} + 2na_n + a_n]x^n = 0.$$

Equating all coefficients to 0, again, leads to the formulas $a_2 = -\frac{a_o}{2}$,

$a_{n+2} = -\frac{(2n+1)}{(n+2)(n+1)} a_n$ for $n \geq 1$. Writing out the first several of these we can determine the general pattern.

$$a_2 = -\frac{a_o}{2}$$

$$a_4 = -\frac{5}{4\cdot 3} a_2 = \frac{(-1)^2 5\cdot 1}{4!} a_o$$

$$a_6 = \frac{-9}{6\cdot 5} a_4 = \frac{(-1)^3 9\cdot 5\cdot 1}{6!} a_o$$

$$a_8 = \frac{-13}{8\cdot 7} a_6 = \frac{(-1)^4 13\cdot 9\cdot 5\cdot 1}{8!} a_o$$

.
.
.

$$a_{2k} = \frac{(-1)^k (4k-3)(4k-7)\ldots 9\cdot 5\cdot 1}{(2k)!} a_o$$

$$a_3 = -\frac{3}{3\cdot 2} a_1 = \frac{(-1)\cdot 3}{3!} a_1$$

$$a_5 = \frac{-7}{5\cdot 4} a_3 = \frac{(-1)^2 7\cdot 3}{5!} a_1$$

$$a_7 = \frac{-11}{7\cdot 6} a_5 = \frac{(-1)^3 11\cdot 7\cdot 3}{7!} a_1$$

$$a_9 = \frac{-15}{9\cdot 8} a_7 = \frac{(-1)^4 15\cdot 11\cdot 7\cdot 3}{9!} a_1$$

.
.
.

$$a_{2k+1} = \frac{(-1)^k (4k-1)(4k-5)\ldots 7\cdot 3}{(2k+1)!} a_1$$

Again a_o and a_1 are arbitrary and we obtain

$$y = a_o[1 + \sum_{k=1}^{\infty} a_{2k}x^{2k}] + a_1[x + \sum_{k=1}^{\infty} a_{2k+1}\ x^{2k+1}]$$

where a_{2k} and a_{2k+1} are given by the above formulas.

7. Let $y = \sum_{n=0}^{\infty} a_n x^n$, $y' = \sum_{n=1}^{\infty} na_n x^{n-1}$, and, rewriting the equation as $xy' + y' + 3y = 0$ and substituting, we obtain:

$$xy' + y' - 3y = \sum_{n=1}^{\infty} na_n x^n + \sum_{n=1}^{\infty} na_n x^{n-1} - 3\sum_{n=0}^{\infty} a_n x^n = 0.$$

In the second series we replace n-1 by n to obtain:

$$\sum_{n=1}^{\infty} na_n x^n + \sum_{n=0}^{\infty} (n+1)a_{n+1}x^n - \sum_{n=0}^{\infty} 3a_n x^n = 0.$$

Now, we write out the $n = 0$ terms of the 2nd and 3rd series and then combine all 3 remaining series which begin at $n = 1$.

$$(a_1 - 3a_o) + \sum_{n=1}^{\infty} [(n+1)a_{n+1} + (n-3)a_n]x^n = 0.$$

Equating all coefficients to zero, we obtain $a_1 = 3a_o$ and, for $n \geq 1$, $a_{n+1} = \frac{3-n}{n+1} a_n$. We now calculate the first several coefficients to see if the general formula can be obtained.

$a_2 = \frac{3-1}{2} a_1 = a_1 = 3a_o$ $\qquad$ $a_3 = \frac{3-2}{3} a_2 = \frac{1}{3} a_2 = \frac{1}{3}(3a_o) = a_o$

$a_4 = \frac{3-3}{4} a_3 = 0$, and now, since every coefficient is a multiple of the preceding one, all of the remaining coefficients are 0. Thus, the solution is: $y = a_o + a_1x + a_2x^2 + a_3x^3 = a_o + 3a_ox + 3a_ox^2 + a_ox^3 = a_o(1+3x+3x^2+x^3) = a_o(1+x)^3$.

10. Substituting the usual power series for y, y' and y'' into the equation we obtain $y'' - xy = \sum_{n=2}^{\infty} n(n-1)a_n x^{n-2} - \sum_{n=0}^{\infty} a_n x^{n+1} = x^4$

$$\Longrightarrow \sum_{n=0}^{\infty} (n+2)(n+1)a_{n+2}\ x^n - \sum_{n=1}^{\infty} a_{n-1}\ x^n = x^4$$

$$\Longrightarrow 2a_2 + \sum_{n=1}^{\infty} [(n+2)(n+1)a_{n+2} - a_{n-1}]x^n = x^4.$$

All coefficients on the left side must be 0 except that of x^4 which must be 1. Thus, with $n = 4$, we obtain $6 \cdot 5\ a_6 - a_3 = 1$, or $a_6 = \frac{a_3+1}{6 \cdot 5}$. In addition, we obtain $a_2 = 0$ and, for all $n \geq 1$ (except $n = 4$) $[(n+2)(n+1)a_{n+2} - a_{n-1}] = 0$

$\Longrightarrow a_{n+2} = \dfrac{a_{n-1}}{(n+2)(n+1)}$. From this formula and $a_2 = 0$ it follows that $a_5 = a_8 = a_{11} = \ldots = a_{3k+2} = \ldots = 0$. We will now write out the first several of the other coefficients to see the patterns.

$$a_3 = \frac{a_o}{3\cdot 2}$$

$$a_6 = \frac{a_3+1}{6\cdot 5} = \frac{a_o}{6\cdot 5\cdot 3\cdot 2} + \frac{1}{6\cdot 5} = \frac{1}{6\cdot 5\cdot 3\cdot 2}(a_o+6) = \frac{4}{6!}(a_o+6)$$

$$a_9 = \frac{a_6}{9\cdot 8} = \frac{4}{9\cdot 8\cdot 6!}(a_o+6) = \frac{7\cdot 4}{9!}(a_o+6)$$

.

.

.

$$a_{3k} = \frac{(3k-2)(3k-5)\ldots 7\cdot 4}{(3k)!}(a_o+6) \text{ for } k \geq 2.$$

Next, $a_4 = \dfrac{a_1}{4\cdot 3} = \dfrac{2}{4!}a_1$. $a_7 = \dfrac{a_4}{7\cdot 6} = \dfrac{5\cdot 2}{7!}a_1$. $a_{10} = \dfrac{a_7}{10\cdot 9} = \dfrac{8\cdot 5\cdot 2}{10!}a_1$, and in general, $a_{3k+1} = \dfrac{(3k-1)(3k-4)\ldots 5\cdot 2}{(3k+1)!}a_1$.

Thus $y = \sum_{k=0}^{\infty} a_{3k+1}x^{3k+1} + a_o + \dfrac{a_o}{6}x^3 + \sum_{k=2}^{\infty} a_{3k}x^{3k}$, where a_{3k+1} and a_{3k} are given by the formulas derived above.

EXERCISES 19.9 (Review), page 998

1. Rewriting as $\dfrac{x}{\csc x}dx - e^{-y}dy = x\sin x\,dx - e^{-y}\,dy = 0$, it is separable. Integrating yields the solution $\sin x - x\cos x + e^{-y} = C$.

4. This is 1st order and linear with integrating factor $e^{\int P} = e^{\int 4} = e^{4x}$. Multiplying by this factor we obtain $e^{4x}y' + 4e^{4x}y = \dfrac{d}{dx}(e^{4x}y) = e^{3x} \Longrightarrow$ $e^{4x}y = \dfrac{1}{3}e^{3x} + C \Longrightarrow y = \dfrac{1}{3}e^{-x} + Ce^{-4x}$

7. This is 1st order and linear with $P(x) = \tan x$ so that $\int P = \int \dfrac{\sin x}{\cos x}dx =$ $-\ln\cos x = \ln\sec x$ and the integrating factor is $\sec x$. Multiplying by this factor we obtain $(\sec x)y' + (\sec x\tan x)y = \dfrac{d}{dx}(\sec x)y = 2\sec^2 x \Longrightarrow$ $(\sec x)y = 2\tan x + C \Longrightarrow y = \dfrac{2\tan x}{\sec x} + \dfrac{C}{\sec x} = 2\sin x + C\cos x$.

10. Rewriting as $(2y+x^3)-xy' = 0$ or $y' - \dfrac{2}{x}y = x^2$, the equation is linear with

integrating factor $e^{\int(-\frac{2}{x})} = e^{-2 \ln x} = 1/x^2$. As usual, we multiply by this factor to get $\frac{1}{x^2} y' = \frac{2}{x^3} y = \frac{d}{dx}(\frac{1}{x^2} y) = 1 \Rightarrow \frac{1}{x^2} y = x + C \Rightarrow y = x^3 + Cx^2$.

13. This is the Bernoulli equation with n = 3. Thus let $w = y^{-2}$ or $y = w^{-1/2}$ so that $y' = -\frac{1}{2} w^{-3/2} w'$. Substituting, $xy' - 2y = x^3y^3$ becomes $-\frac{1}{2} xw^{-3/2}w' - 2 w^{-1/2} = x^3w^{-3/2}$. Multiplying through by $w^{3/2}$ we obtain the linear equation in w: $-\frac{1}{2} xw' - 2w = x^3$ or $w' + \frac{4}{x} w = -2x^2$. The integrating factor is $e^{\int 4/x} = e^{4 \ln x} = x^4$ and thus we obtain $\frac{d}{dx}(x^4w) = -2x^6 \Rightarrow x^4w = -\frac{2}{7} x^7 + C \Rightarrow$ $w = y^{-2} = -\frac{2}{7} x^3 + \frac{C}{x^4}$. Explicitly, $y = (-\frac{2}{7} x^3 + \frac{C}{x^4})^{-1/2}$.

16. $m^2-6m+25 = 0 \Rightarrow m = \frac{6 \pm \sqrt{36-100}}{2} = \frac{6 \pm 8i}{2} = 3 \pm 4i$. Thus $y = e^{3x}(C_1 \cos 4x + C_2 \sin 4x)$.

19. The complementary equation, $y''-y = 0$, has auxiliary equation $m^2-1 = 0$ with solutions $m = \pm 1$. The right side of the given equation, $e^x \sin x$, has s = t = 1 in the notation of 19.36(iii). Since the complex number 1+i is not a solution of the auxiliary equation, the given differential equation has a particular solution of the form $y_p = e^x(A \cos x + B \sin x)$. Computing $y_p' = e^x((A-B)\sin x + (A+B)\cos x)$, $y_p'' = e^x(-2B \sin x + 2A \cos x)$, and substituting into the left side of the given equation we end up with $y_p'' - y_p = e^x[(-2B-A)\sin x + (2A-B)\cos x]$. This will equal $e^x \sin x$ if $-2B-A = 1$, $2A-B = 0$. From the 2nd, B = 2A, and from the 1st we get A = -1/5, B = -2/5. Thus $y_p = e^x(-\frac{1}{5} \cos x - \frac{2}{5} \sin x)$ and the general solution is $y = C_1e^x + C_2e^{-x} + e^x(-\frac{1}{5} \cos x - \frac{2}{5} \sin x)$.

22. This equation is exact since $P_y = Q_x = -\frac{1}{x} + \frac{1}{y}$. Thus we seek F(x,y) such that $F_x = P = 2x-yx^{-1} + \ln y$, $F_y = Q = xy^{-1} - \ln x + 1$. From the 1st, $F(x,y) = x^2 - y \ln x + x \ln y + g(y) \Rightarrow F_y = -\ln x + xy^{-1} + g'(y)$. Comparing with the F_y equation we find $g'(y) = 1$ or $g(y) = y + C_1$, and the solution is $x^2 - y \ln x + x \ln y + y = C$.

25. The complementary equation, $y'' - 3y' + 2y = 0$ has auxiliary equation $m^2-3m+2 = (m-2)(m-1) = 0$, with solutions m = 1,2. The right side of the given equation is e^{5x}, and since 5 is not a solution of the above auxiliary equation, there is a particular solution $y_p = Ae^{5x} \Rightarrow y_p' = 5A e^{5x} \Rightarrow$

$y_p'' = 25A\, e^{5x}$. Substituting into the equation: $y_p'' - 3y_p' + 2y_p = (25A - 15A + 2A)e^{5x} = 12A\, e^{5x} = e^{5x}$ if $12A = 1$ or $A = 1/12$. Thus $y_p = 1/12\, e^{5x}$ and the general solution is $y = C_1e^x + C_2e^{2x} + \frac{1}{12}e^{5x}$.

28. This equation is exact since $P_y = Q_x = -4xy$. Thus we seek $F(x,y)$ such that $F_x = P = 3x^2 - 2xy^2 + 1$ and $F_y = Q = y^2 - 2x^2y$. From the 1st $F(x,y) = x^3 - x^2y^2 + x + g(y) \Rightarrow F_y = -2x^2y + g'(y)$. Thus $g'(y) = y^2 \Rightarrow g(y) = \frac{y^3}{3} + C_1$ and the solution is $x^3 - x^2y^2 + x + \frac{y^3}{3} = C$.

31. $m^2 + 5m + 7 = 0 \Rightarrow m = \frac{-5 \pm \sqrt{25-28}}{2} = -\frac{5}{2} \pm \frac{\sqrt{3}}{2}i \Rightarrow y = e^{-\frac{5}{2}x}(C_1\cos\frac{\sqrt{3}}{2}x + C_2\sin\frac{\sqrt{3}}{2}x)$.

34. $m^2 + 10m + 25 = (m+5)^2 = 0 \Rightarrow -5$ is a double root $\Rightarrow y = (C_1 + C_2x)e^{-5x}$.

37. The equation is exact since $P_y = Q_x = 1 - 2(\cos y)e^{-2x}$. Thus once again we seek $F(x,y)$ such that $F_x = P = y - 2(\sin y)e^{-2x}$, $F_y = x + e^{-2x}\cos y$. From the 1st, $F(x,y) = xy + (\sin y)e^{-2x} + g(y) \Rightarrow F_y = x + (\cos y)e^{-2x} + g'(y)$ $\Rightarrow g'(y) = 0$ and the solution is $xy + (\sin y)e^{-2x} = C$.

40. This last equation is of homogeneous type. We'll rewrite it as $(x^3+y^3)dx - xy^2dy = 0$ and let $y = vx$, $dy = v\,dx + x\,dv$ to obtain $(x^3 + v^3x^3)dx - xv^2x^2(v\,dx + x\,dv) = 0 \Rightarrow x^3(1 + v^3-v^3)dx - x^4v^2dv = 0 \Rightarrow \frac{dx}{x} - v^2dv = 0 \Rightarrow \ln|x| - \frac{v^3}{3} = C \Rightarrow \ln|x| - \frac{y^3}{3x^3} = C$.